The Speed of Air

The Story of Willard Custer and His Channel Wing Aircraft

by

Joel C. Custer and Robert J. Englar

THE SPEED OF AIR

The Story of Willard Custer and his Channel Wing Aircraft.

Foreword, Willard R. Custer as the Father of Powered Lift Aircraft.

Afterword, Where Do We Go from Here? The Future of the Channel Wing.

Published by Four-Thirty Press, LLC, Ashburn, VA.

Cover design by Jonathan Ameen.

Additional graphics by Hana Massalski.

Library of Congress Control Number: 2023907451

To Sara Creeger

Willard R. Custer (1899-1985)

Contents

Foreword i

Author's Note xiii

Chapter 1, The Speed of Air 1925 - 1940 1

Chapter 2, The Bumblebee 1940 - 1943 17

Chapter 3, The Experts 1944 - 1948 35

Chapter 4, The Experiment 1949 - 1953 59

Chapter 5, The Prototype 1953 - 1957 77

Chapter 6, The Stockholders 1958 - 1963 103

Chapter 7, The Production Model 1964 - 1973 129

Epilogue 151

Afterword 161

Acknowledgements 189

Notes 193

Bibliography 215

Index 229

About the Authors 233

The credit belongs to the man who is actually in the arena,
whose face is marred by dust and sweat and blood;
who strives valiantly;
who errs, who comes short again and again,
because there is no effort without error and shortcoming;
but who does actually strive to do the deeds;
who knows great enthusiasms, the great devotions;
who spends himself in a worthy cause;
who at the best knows in the end the triumph of high achievement,
and who at the worst, if he fails, at least fails while daring greatly,
so that his place shall never be with those cold and timid souls who
neither know victory nor defeat.

-- Theodore Roosevelt, Paris, 1910

Foreword

Willard R. Custer as the Father of Powered Lift Aircraft

by

Robert J. Englar

Today's transportation industry depends on new technology development to make potential changes, often leading to either future financial successes or possible downturns. The aviation/aerospace segment, whether commercial or military, depends on these new innovative technologies to make major leaps in performance, operational advances, and financial paybacks. This very advancement of new technology indeed became the early goal of Maryland mechanic/inventor Willard R. Custer, who by training was neither an aerodynamicist nor an aircraft designer. In the 1927-1942 timeframe, Custer developed and successfully flew a new and very novel aircraft concept, covered by over 25 patents, that eventually became known as the Custer Channel Wing.

This early "Powered Lift" non-helicopter winged aircraft achieved some amazing tasks. These included taking off vertically (hovering) and flying at very low air speeds, all with virtually no moving aerodynamic parts besides a propeller. It thus overcame a lift-generating speed requirement (where the lift force generated on an aircraft normally increases with the square of its flight speed) by using the engine's output thrust stream to provide extra lift. This eliminated the complex mechanical components of conventional high-lift systems. The early-demonstrated payoffs of his prototype included: no runway required and no rotating blades overhead (during vertical takeoff/landing), and/or very short runways required and resulting greater liftable payloads, increased airport safety, and reduced noise (resulting from very low required liftoff/touchdown flight speeds). Applied to some of today's transportation issues, this could result in personal air vehicles that could takeoff and land from your own driveway; smaller/simpler naval aircraft carriers without mechanical catapults or arresting gear; and small regional commercial airports accepting larger aircraft without longer runways or noise issues.

However, although Custer successfully demonstrated his new Powered Lift concept's abilities to a wide range of observers during flight tests of five different Channel Wing aircraft versions, he was not able to acquire the support/backing he needed to make his Channel Wing into a saleable and operational commercial aircraft. So why wasn't the Channel Wing aircraft a commercial success? Basically, Custer was pitted against both government scientists/evaluators and commercial aircraft builders/operators, many claiming his aircraft were too unconventional to succeed, or that his salesmanship efforts were too overwhelming. **That history and these obstacles** will all be revealed in the following chapters. Nevertheless, he pressed on from 1927 into the 1980's to develop and prove his Channel Wing concept to the point where a "Production" version was flying. Although Custer never saw his dream of a commercial operational Chanel Wing come true in his lifetime, these efforts by Custer have now led to more recent Channel Wing developments by other investigators/researchers (including this co-author), and are now being considered for incorporation into future Powered Lift aircraft designs.

Before we discuss the multiple decades of development and trials/tribulations of Willard Custer's Channel Wing aircraft, it would be useful for the reader to first understand the basic problem of low-speed aircraft takeoff and landing operations, and why a significant improvement in aircraft design such as the Powered Lift concept incorporated in the Channel Wing could offer great promise in future aviation systems. Ever since fixed-wing aircraft (not helicopters) have operated from hard runways or carrier decks, much importance has been placed on their takeoff and landing ("terminal area") capability. Very desirable here is the aircraft's ability to increase the available lift coefficient (C_L) to well above that value employed in higher-speed cruise flight. This C_L increase will decrease the flight speed needed to support the aircraft's weight in its steady approach to landing or in its liftoff. It will also reduce the landing or takeoff ground roll distances. On conventional aircraft wings, this increase in the available lift is usually obtained by one of the following: cambering (curving down) the wing's trailing edge by use of mechanical flaps; increasing the aircraft's angle of attack to the air; or using leading-edge devices such as slats to increase C_{Lmax}

by delaying the stall angle of attack at which the airflow separates. In more advanced designs, the lift from the flaps can be further augmented by use of small blowing jets/slots which keep the separating flow attached to higher flap angles. In all of these cases, the attached airflow causes higher local velocities and thus lower static pressures (suction) on the wing upper surface plus increased pressure on the lower surface, and thus significant increases in the available lift.

In the extreme, the aircraft's propulsion unit(s) can be used to greatly increase the local velocity over the wing to values much higher than the flight speed of the aircraft through the air (the airspeed). This has become known as **Powered Lift**. This book will show that **Powered Lift was initially conceived by Maryland mechanic/inventor Willard R. Custer** and was demonstrated on a pilotless powered model as early as the year 1927. Custer further developed the technology in experimental and flight test efforts on his Custer Channel Wing (CCW) aircraft, with the first manned CCW aircraft flying over 100 flight test hours in 1942.

A specific class of fixed-wing aircraft has been defined by the Federal Aviation Authority (FAA) in its Federal Aviation Regulations (FAR). The Powered-Lift aircraft category "means a heavier-than-air aircraft capable of vertical takeoff, vertical landing, and low-speed flight that depends on engine-driven lift devices or engine thrust for lift during these flight regimes, and on non-rotating airfoil(s) for lift during horizontal flight" [Ref. 1]. Note that this definition requires a vertical takeoff and landing capability, but it eliminates helicopters and all rotary-wing aircraft from this category.

In 1927, long before these characteristics were prescribed by the FAA, Willard Custer had realized from his observations of weather over buildings, and from his early model experiments, that the critical factor for lift on these types of aircraft was not so much the ***aircraft's speed through the air, but rather the speed of the local airflow over the aircraft's lifting surface(s)***, nominally the wing. If local velocity could be induced or entrained over the wing's upper surface, then corresponding vertical lift could be generated on the aircraft, ***even if it had no forward speed***. This is Vertical Takeoff and Landing (VTOL), with no runway length required. Experimenting on a model with internal combustion engines driving conventional aircraft propellers, Custer confirmed that if the propeller's slipstream velocity were enclosed within a trough (a channel, Figure F.1), increased velocities were created by that slipstream on the trough's upper surface. These velocities were sufficient (because of the inverse relationship between local velocity and static pressure) to cause the trough to rise vertically, perpendicular to the slipstream direction. This was the earliest version of the channel

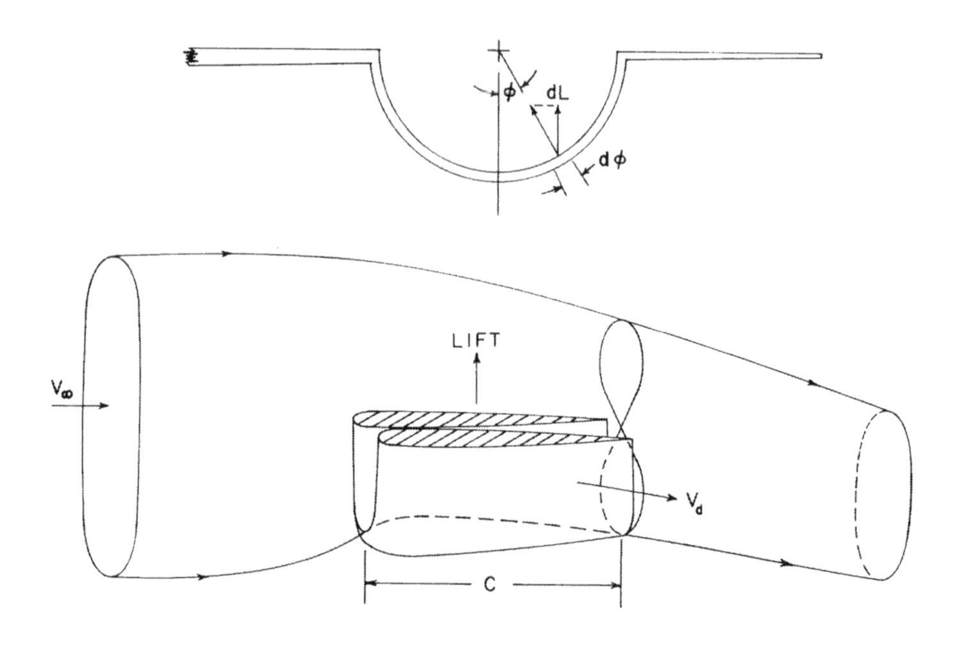

Figure F.1 — Channel, propeller, and prop slipstream

wing concept, and the Custer Channel Wing (CCW) was thus born in 1927. If this concept were employed in an aircraft, it would meet the FAA Powered Lift requirements to provide vertical lift and VTOL. It could also provide Short Takeoff and Landing (STOL), assuming that factors such as low-speed control were also met. At this point, the reader might easily conclude that Custer may well have been the earliest developer of Powered Lift for these aeronautical purposes.

There have been many Powered Lift aircraft developed and flown since that time to provide alternatives to the conventional rotary-winged helicopter, which is typically limited in its maximum forward speed by stall of the retreating rotor blade. These aircraft all employ their primary propulsive unit(s) or auxiliary power units as a source of increased velocity or engine downward thrust to enable vertical lift/flight capability. Some examples [Ref. 2], including their VTOL types of propulsive/lift units and their first flight dates, are as follows (a few of which are STOL):

Aircraft	*Propulsion/ Lift Type*	*First Flight*
LTV XC-142 (Figure F.4)	*Tilt prop/engine (4) on tilt wing*	*1964*
Canadair CL-84	*Tilt prop/engine (2) on tilt wing*	*1965*
Dassault Balzar Mirage III-V	*8 lift jet, mach 2 fighter, fixed wing*	*1965*
Bell X-22 (Figure F.3)	*Tilt ducted fan (4), fixed wing*	*1966*
Dornier Do 31 (Figure F.5)	*8 embedded vertical lift jets, fixed wing transport*	*1967*
Curtis Wright X-19 (Figure F.2)	*Tilt rotor (4), fixed wing*	*1977*
McDonnell Douglas AV-8B Harrier (Figure F.9)	*Vectored thrust nozzles (4) fixed wing with flaps*	*1977*
Bell XV-15	*Tilt rotor (2), fixed wing*	*1977*
Rockwell XFV-12A (Figure F.8)	*Augmentor wing V/STOL fighter, fixed wing*	*1977*
NASA QSRA	*Upper surface blowing (4 engines) over flap, STOL*	*1978*
Yakolev Yak-141	*Vectored thrust nozzles, supersonic fighter*	*1987*
Bell Boeing V-22 Osprey	*Tilt rotor (2), fixed wing*	*1989*
Bell Augusta AW609	*Civilian tilt rotor (2), fixed wing*	*1996*
Lockheed F-35B Lightning II (Figure F.10)	*Vectored thrust + lift engine, STOVL fighter*	*2006*

Clearly Custer's Channel Wing, with his earliest manned version flying in 1942 (the unmanned version flew in 1927), flew earlier than any of the powered-lift aircraft listed above. **This timeframe leads to Willard R. Custer qualifying as the initiator of the first aircraft in the Powered-Lift aircraft category, and thus earning the title "The Father of Powered Lift."**

A few comparisons of the Custer Channel Wing aircraft with the above powered-lift aircraft are in order here. Most of those listed above (and this is by no means a complete historical listing) were intended to provide VTOL by using a number of differing engine arrangements.

To achieve the vertical "lift" component required for either VTOL or STOL takeoff and landing (in many cases, this was just tilted thrust vector-

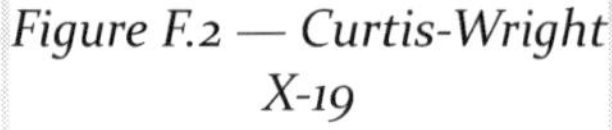

Figure F.2 — Curtis-Wright X-19

Figure F.3 — Bell X-22

Figure F.4 — LTV XC-142A

ing with little aerodynamic lift component), many of these aircraft have employed mechanically-tilting rotors or props (Figures F.2, F.3), or tilting wing/prop/engine combinations (Figure F.4). Others used multiple jet engines (Figure F.5) or lift fans (Figure F.6) or ejectors (Figures F.7, F.8) deployed vertically within the fuselage, wings, or pods, all of which needed to be retracted or covered over for cruise flight.

Still others used mechanically deflecting jet exhaust nozzles (Figures F.9, F.10), or were tail sitters (Figure F.11). Achieving vertical lift from the engines alone was always a serious propulsion design issue because the engine(s) had to provide enough vertical thrust to yield a thrust-to-weight ratio of 1.0 or greater.

In many cases, some of these became combined Vertical/Short Takeoff and Landing (V/STOL) or Short Takeoff/Vertical Landing, (STOVL), operating in the Short mode only for takeoff (the heaviest gross weight) and in the Vertical mode only at a much lower landing weight. Plus, any design that had any **moving/tilting** rotors or wings or nozzles or ducts or fans or vanes or auxiliary engines would then also add extraneous factors: much mechanical complexity (and less reliability); greatly increased weight (and

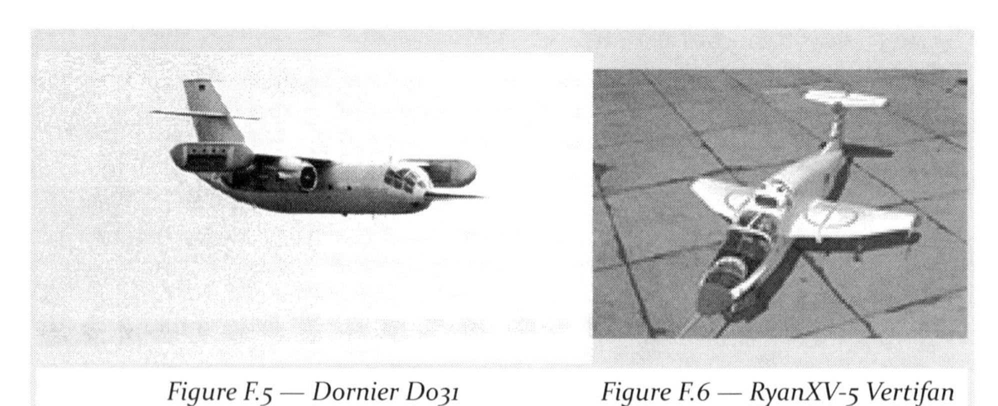

Figure F.5 — Dornier Do31

Figure F.6 — RyanXV-5 Vertifan

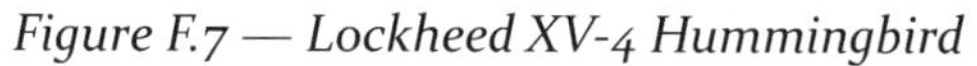
Figure F.7 — Lockheed XV-4 Hummingbird

Figure F.8 — Rockwell XFV-12A Augmentor Wing

thus performance loss); reduced fuel efficiency (due to oversized or extra engines); and thus in all cases, higher development, procurement and operational costs.

Amazingly, not only did the Custer Channel Wing aircraft avoid all these issues, it did so in the simplest possible manner to obtain vertical or short flight operations. It had no moving components besides the conventional rotating props, internal engine parts, and aerodynamic control surfaces. It achieved variation in the thrust-induced lift by varying prop RPM (and the thrust coefficient, C_T) and could increase the thrust lift component ($C_T \sin \alpha$) by increasing aircraft angle of attack, α.

So, how much powered-lift could the Channel Wing generate? Figure F.12 shows flight test data for Custer's CCW-5 aircraft, where lift coefficient

Figure F.9 — McDonnell Douglas AV-8B Harrier

Figure F.10 — Lockheed F-35B Lightning II

Figure F.11 — Lockheed XFV-1 Salmon

C_L increased when angle of attack α and thrust coefficient C_T were increased. Non-dimensional C_L = lift force / (wing area x flight dynamic pressure). The thrust coefficient C_T is total engine thrust divided by (wing area x flight dynamic pressure). The dashed lower curve labeled Power Off is the lift due only to the unpowered channels and wing, and is quite typical to a conventional fixed-wing aircraft: maximum C_L of around 1.5 and stall angle of about α= 24 deg (a bit higher than conventional aircraft). The circle symbols are

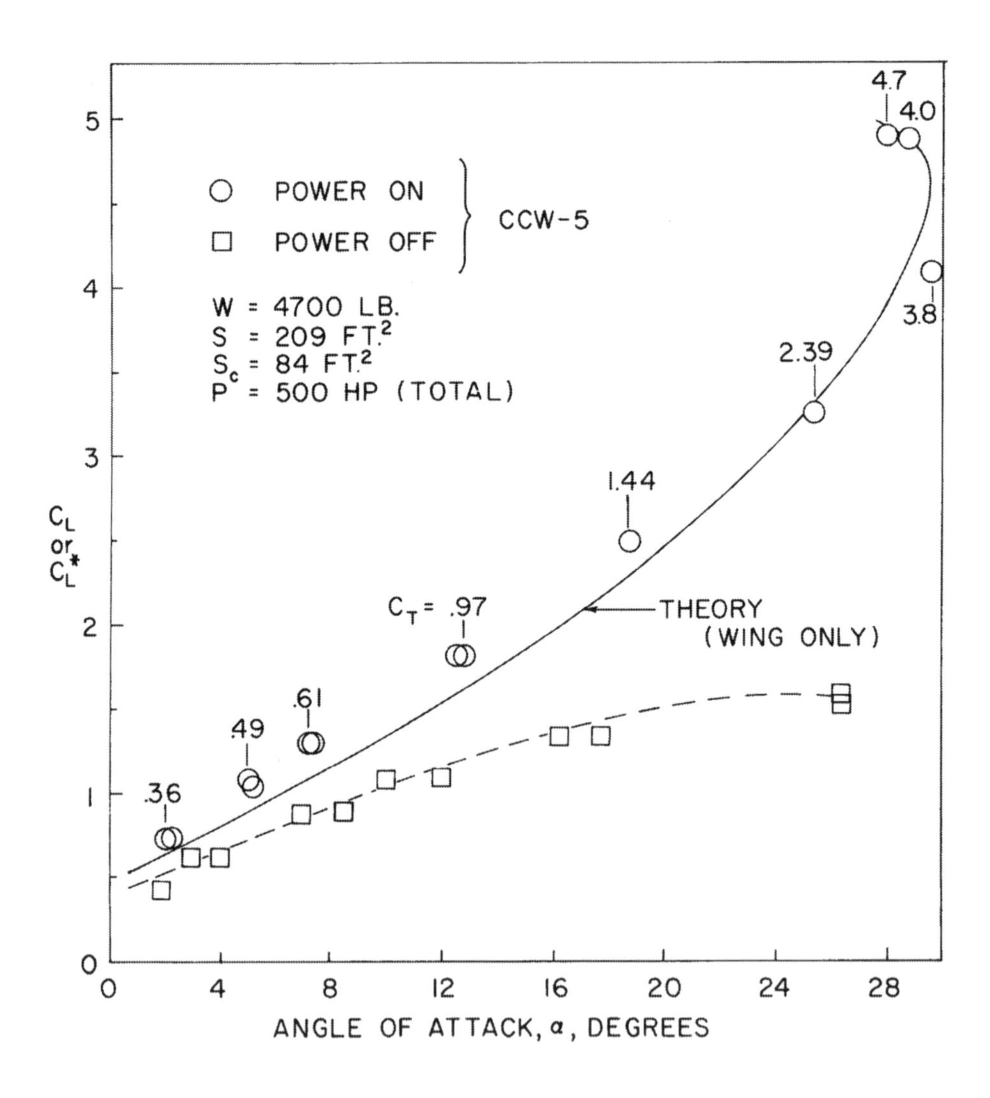

Figure F.12 —CCW Flight test data

for the CCW-5 with Power On and C_T greater than 0. Note that as C_T and α are increased for the CCW, the lift coefficient increases from 0.7 to nearly 4.8, a 220% increase of the Power Off maximum lift. The difference between the two curves represents the CCW-generated Powered Lift. Thus the operating mode for both takeoff and landing by CCW can be very similar to a conventional fixed-wing aircraft, except for the great increase in CCW available lift and the vertical/STOL flight possibility.

The earliest manned Custer Channel Wing flight vehicle flew over 100 hours of test flights in 1942 [Ref. 3]. On the calendar chart above, early flying VTOL aircraft entries were the Bell Model 65 (2 tilting turbojet engines below the wing) and the Lockheed XFV-1 (Figure F.11, a turboprop fixed-wing twin prop tail-sitter), both with first flight in 1954; the Bell XV-3 tilt rotor/fixed wing in 1955, and the Boeing Vertol VZ-2 tilt wing/prop/engine in 1957. These aircraft employed the mechanical and structural complexity of tilting either the rotors, the engines, or the wing/engine/prop combination to obtain vertical thrust for VTOL, or to takeoff/land vertically as a tailsitter with the related operational difficulties. The earliest manned flying versions of these (which were probably more complex and costly than CCW) were 12 or more years later than Custer's first flying powered-lift manned aircraft.

It is interesting to note that the only other flying aircraft similar to the Custer Channel Wing was the German Rhein Flugzeugbau FR-1 (not pictured). This had a single 180-degree channel with prop located directly behind the fuselage, but also had a 180-degree shroud on top of the channel, so that it really was a ducted-fan configuration. This ducted fan may well have augmented the engine thrust component, but the negative static pressure inside that top shroud probably offset much of the suction in the lower channel as well as its vertical lift vector. Its first flight was in 1960, 18 years later than Custer's CCW. A Russian channel wing aircraft (the Antonov 181, not pictured) was built in the late 1980s (after Custer's death), but it never flew; it achieved only high-speed taxis. This might well have been because the prop was located at the very front of the channel, yielding a higher static pressure field behind the prop and in the channel, which probably reduced any lift augmentation. This prop location was a CCW design lesson that Willard had already discovered many years earlier, but the Russians must not have!

Associated with the title of **Father of Powered Lift Aircraft** should be the basic premise of the Custer Channel Wing: In VTOL and STOL Powered Lift, what is critical is **the speed of the air** over the lifting surfaces (for CCW, this is the entrained air velocity being pulled over the channel surface into the aft propeller, Figure F.1) **rather than the speed of the aircraft** into the ambient air. For VTOL aircraft, this windspeed is by definition **zero**, unless there is a sidewind, which then becomes a flight control issue. This is also true for STOL, V/STOL or STOVL aircraft, where the low flight speed needed for short takeoff and landing distances would normally not provide the wind dynamic pressure and lift force necessary to support the aircraft without the power component. STOL or Super STOL flight can be enhanced when the **speed of the air entrained into the lifting surface of the channel wing by the prop can be combined with the STOL flight speed (including any head wind) to generate the necessary lift.**

The following chapters will certify these claims for Willard Custer's Channel Wing Aircraft and will detail the trail that Custer blazed to embody his CCW design into a family of successful aircraft. They will also answer the question that always occurs to every serious aviation enthusiast who reviews the Channel Wing's history and achievements: "Whatever happened to it, and why isn't it flying now?"

The story of the Channel Wing doesn't end with the Custer CCW-5 Production aircraft (1964-1973). Interest in this Powered Lift aircraft technology has continued [Refs. 4,5,6]. New developments and associated improvements have occurred, and new versions/configurations of the Channel Wing Powered Lift Aircraft are presented in the Afterword. Resolution of some of the operational issues for this type of Powered-Lift VTOL/STOL aircraft plus additional new features and capabilities of advanced Channel Wing Powered Lift Aircraft will also be addressed in that chapter.

References

[1] Mattingly, Daniel, "Flight Control Design Characteristics of a Civilian Powered Lift Category Aircraft," Southern Illinois University, Dept. of Aviation Technologies, 2011.

[2] "STOVL/VTOL Aircraft Ranked," www.oppositelock.kinja.com.

[3] Blick, E. F., "The Channel Wing - An Answer to the STOL Problem," Shell Aviation News No. 392, 2-7, 1971, June 27, 2017.

[4] Englar, Robert J. and Brian A. Campbell, "Development of Pneumatic Channel Wing Powered-Lift Advanced Super-STOL Aircraft," AIAA Paper 2002-2929, AIAA 20th Applied Aerodynamics Conference, St. Louis, MO, June 25, 2002.

[5] Englar, R. J. and B. A. Campbell, "Pneumatic Channel Wing Powered-Lift Advanced Super-STOL Aircraft," AIAA Paper 2002-3275, AIAA 1st Flow Control Conference, St. Louis, MO, June 26, 2002.

[6] Englar, Robert J. & Brian. A. Campbell, "Experimental Development and Evaluation of Pneumatic Powered-Lift Super-STOL Aircraft", Paper #3 presented at the NASA/ONR Circulation Control Workshop, Hampton, VA, March 2004. Also published in "Workshop Proceedings", NASA CP 2005-213509, 2005.

Photo Attribution

F.2 – Public Domain, fantastic-plastic.com/CurtisWrightX-19-Attic.htm

F.3 – (substituted) Public Domain, commons.wikimedia.org/wiki/File:X-22a_onground_bw.jpg

F.4 – (substituted) By NASA - http://lisar.larc.nasa.gov/UTILS/info.cgi?id=EL-2001-00399, Public Domain, https://commons.wikimedia.org/w/index.php?curid=12486606 (color removed)

F.5 – Per License, commons.wikimedia.org/wiki/File:Dornier_Do_31_in_1968.jpg (color removed)

F.6 – (substituted) Public Domain, en.wikipedia.org/wiki/Ryan_XV-5_Vertifan (color removed)

F.7 – Public Domain, commons.wikimedia.org/w/index.php?curid=3376949 (color removed)

F.8 – Public Domain, commons.wikimedia.org/wiki/File:Rockwell_XFV-12_hovering_over_ship_-_drawing_1974.jpg (color removed)

F.9 – (substituted) Public Domain, commons.wikimedia.org/wiki/File:2012_

MCAS_Miramar_AirShow_121013-M-GC438-375.jpg (color removed)

F.10 – (substituted) By U.S. Navy photo by Mass Communication Specialist 3rd Class Nathan T. Beard - This image was released by the United States Navy with the ID 191017-N-QI061-1159 Public Domain, commons.wikimedia.org/w/index.php?curid=83394720 (color removed)

F.11 – (substituted) Public Domain, commons.wikimedia.org/wiki/File:Lockheed_XFV-1_on_ground_bw.jpg

Author's Note

We were walking together across Sally Buffalo Park in the little town of Cadiz, Ohio, just ten miles south of New Rumley, the birthplace of George Armstrong Custer. In fact, we were talking about General Custer. After all, the occasion was "Custer Day," held every June in New Rumley to commemorate the general, his life, and times. I was walking with Dr. Tom Crouch, senior curator for the Smithsonian Institution's National Air and Space Museum. He had just finished a presentation about Custer's experience with reconnaissance balloons in the early days of the Civil War. After his presentation I introduced myself and, as we walked toward our cars, I remarked to him that researching the general's Civil War career was a hobby of mine. My granddad had told me that we were related to the general, and I confided to Dr. Crouch that I hoped one day to write a book about some aspect of the general's Civil War experiences.

The memory of my granddad, however, interrupted my train of thought. I inquired whether Dr. Crouch might be familiar with my granddad's airplane. The Smithsonian had acquired it—the first Custer Channel Wing (CCW-1)—in 1961. With a look of recognition, Dr. Crouch immediately responded, "Of course, I'm familiar with the Channel Wing! And you are Willard Custer's grandson?" I nodded. "Well, if you're going to write a book, that's the story you should tell."

In 1977, Walt Boyne, a former director of the National Air and Space Museum and a best-selling author, published an article entitled "The Custer Channel Wing Story" in *Airpower* magazine. Since that time, his has been the standard account and the most complete history of Willard Custer and his aircraft in print. Written between Willard's retirement and his death, the article is replete with photographs and crammed with details previously unpublished.

But the story that Boyne tells is still incomplete. Writing for a magazine, his article length was limited. Then he sacrificed some of that space to revisit the engineering debate over the Channel Wing. And Willard—at the time deep in a lawsuit against Fairchild Aircraft Company—kept some of the story to himself. As a result, there are many details and dramatic turns to Willard's story that Boyne could not have known nor had space to reveal. In writing this book, I have no such limitations.

I choose to focus primarily on the people, issues, and events of the Channel Wing story. Aircraft design issues remain at the story's heart, but being a layman myself, I describe them from a layman's view. This is so the nontechnical reader will understand these issues enough to appreciate their significance. For the pilots and aerodynamics engineers in the audience, I gratefully leave the aerodynamics and engineering discussions to the expert, Robert Englar, who ably provides the Foreword and Afterword to this book.

This book is a work of nonfiction based largely on primary documents. While in college, I began collecting pictures, magazines, and corporate handout material on the aircraft. After graduate school, I began researching my grandfather's patents and discussing his concept with aerodynamics experts and entrepreneurs. That research led me to corporation documents, court records, and the National Archives. For years I worked in Washington, D.C., spending my lunch hours doggedly exploring the SEC, Library of Congress, U.S. Patent and Trademark Office, and district court libraries, inspired to dig further with each discovery. I have digested court cases, patent wrappers, and FAA records that other Channel Wing writers have not even scratched.

Then I published my own Channel Wing website, which opened a communications link to the world. I received emails from people in New Zealand, Canada, the United Kingdom, and France, as well as Pennsylvania, New Jersey, and Florida, all with Channel Wing memories, memorabilia, pictures, letters, and stories, offering to send them to me as the most appropriate guardian. As a result, I am able to reveal and describe persons whose involvement in the Custer Channel Wing has never before been disclosed or described in print, and I include photographs that have never been published.

My intent is simply to tell the story as completely, clearly, and impartially as possible, while engaging the reader. What I did not want to write was a history book.

To recreate conversations, I relied on my research and actual events, while taking a few artistic liberties to insert humor and personality. I have documented my sources in the Notes, and I have detailed there how I reconstructed each conversation. The courtroom dialog is taken verbatim, though heavily redacted, from the court transcripts.

This book is not a memoir, and I have endeavored to remain impartial but not impersonal. After all, Willard Custer was my grandfather. I remember sitting on his knee. My first-hand experiences with his aircraft consisted of hours passed in the cockpit of the two CCW-5s, playing at the controls. These aircraft were firmly planted on the ground and in the hangar while my father, Reed, tuned their engines. I was all of seven years old then. And, for a few childhood years thereafter, I occasionally sat at my grandfather's kitchen table or in his living room, shared a Coca-Cola and lemon meringue pie and listened to him tell a few of the stories of the Channel Wing. But because of family circumstances, those visits were not numerous. In my adult years, I've had only a handful of conversations with Willard or my father, and I learned that memories fade. Now the family members of my father's generation have passed away. In whole, my personal experiences with them were regrettably too few. For this book, then, I was able to draw on my own memories of scenes and personalities and only a few anecdotes passed down to me from my family.

Thirty-five years have passed since Dr. Crouch suggested that I write this book. If I must admit to any subjectivity, it is in this one predisposition: I have persevered in this task because the story itself never ceases to inspire me.

Primarily, I continue to be inspired by Willard's independence of thought. Willard conceived of a new wing design and had a confidence in his concept that could not be shaken. It was not based on textbook knowledge or expert opinions, but on his own observations. He knew how things worked, and he relied on his own ability to experiment and to interpret the evidence. In fact, he was so confident in himself that he truly believed that he was right about the Channel Wing—and that the rest of the aircraft industry was wrong. I admire that. In this way, Willard inspires me to think for myself, to seek my own answers, to do my own work, to trust my own interpretations, to follow my own intuitions. He gives me permission to trust my internal compass even if it seems, at times, that my peers are headed in a different direction.

Secondly, I am inspired by Willard's devotion. He demonstrated what it means to never give up. Over his 86-year lifespan he gave all he had in the pursuit of his dream. He is a true-to-life example of complete devotion in the face of long odds and immense challenges.

Finally, this story inspires me to be careful. Desperation can warp the way we see our circumstances. Then we may act—based on faulty perceptions—in ways that run counter to our ideals. Willard did this. He was convinced that the SEC was engaged in a conspiracy to undo him, and so he uncharacteristically defied a District Court judge and suffered for it.

Inspired in this way, my enthusiasm for this book has never flagged. I worked at it in my spare time over thirty-five years. But the more I uncovered the details of this story, the more the story recommended itself to me. My only desire now is to tell it in such a way that my readers will also be inspired.

Joel Custer

In the ancient land of the Manahoac and Monacan People

Loudoun County, Virginia

August 2022

www.custerchannelwing.net

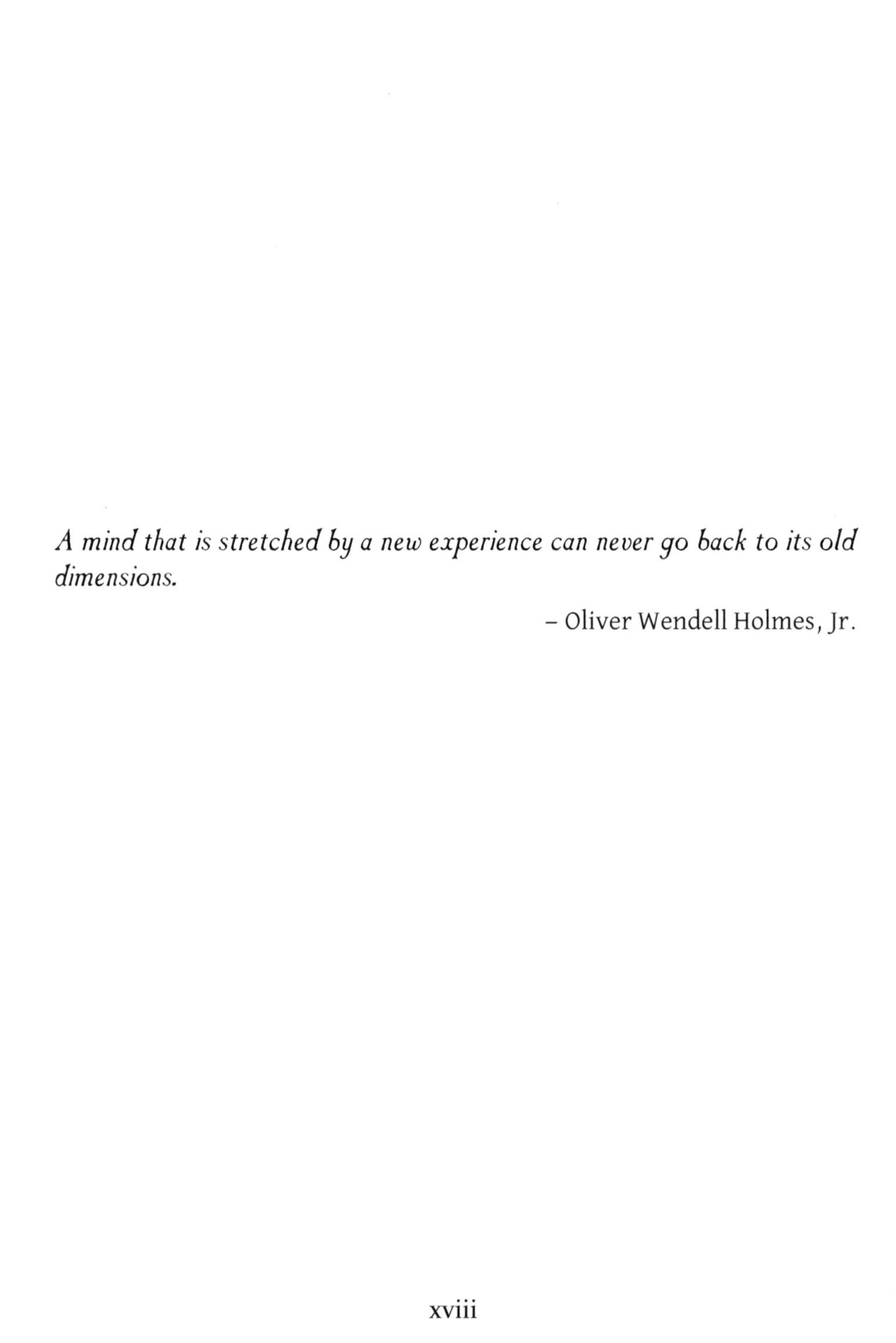

A mind that is stretched by a new experience can never go back to its old dimensions.

– Oliver Wendell Holmes, Jr.

I The Speed of Air 1925 - 1940

A lifting action will be created causing the vessel to move upwardly without traveling forwardly.[1]

- Willard Custer

Willard Custer was a wizard with engines, but the engine on this tractor proved particularly ornery. Normally, he could diagnose an internal combustion ailment as if by a sixth sense. In fact, this skill had earned him his present job as a car mechanic at the Ford Motor Company in Hagerstown, Maryland. This summer afternoon in 1925[2] found him absorbed in such diagnostics on a tractor in a small meadow in Back Creek Valley[3] near Martinsburg, West Virginia, just 25 miles southwest of his Hagerstown home. In shirt and tie underneath his mechanic's overalls, he persisted.

The mechanical mule also distracted him from impending danger. The sun shone clear in the skies above him, but behind him, like a line of charging Civil War cavalry, a front of black-hearted thunderheads quickly approached.

In the summer months, such thunderstorms commonly race head-long out of the Allegheny Mountains, gallop eastward across twenty miles of foothills and the expansive floor of the Cumberland Valley, and then slam into the mountainous ridge that forms the eastern wall. Their reckless charges across the valley are ferocious, and their effects are unpredictable and sometimes indelible. These storms drive a front line of gray clouds like a bulldozer blade across the blue sky, followed almost immediately by a silver drape of drenching rain. Next, their lightning may set a house or barn ablaze, or the torrent may flood the stream beds in a flash. Other times their winds target a long-standing tree or outbuilding with sniper precision and instantly remove it.

Such a storm accelerated toward 26-year-old Willard, who remained totally unaware of its impending sweep of violence. The trees and rolling hills surrounding him shielded his view of the blackening horizon, while the decibels put out by the obstinate tractor engine nearly masked the guttural growl of approaching thunder.

Dousing the sun, the monster sprang. Hot lightning shot overhead. Thunder pounded the earth. The onrush of wind swept the landscape, sending trees into a panic. Alarmed, Willard abandoned his wrenches and dashed to the nearby barn for refuge just ahead of what looked like Niagara Falls rolling his direction.

Once inside, he barred the door against the violent gusts. He removed his hat, loosened his tie, and caught his breath. He looked around for a seat to wait out the storm. Outside, the wind and water surged like a pounding surf against the wall and roof. He was thinking that a good dousing in the rain might actually feel refreshing when a loud, explosive crack yanked his gaze upward. Was the roof collapsing?

He dove to the floor at the base of the nearest wall just as the roof tore loose from the timber posts and catapulted upward into the churning, black clouds. He scurried on hands and knees to the nearest corner and pressed himself into it. Like a wild rabbit, hunted and cornered, he sat motionless, his brown eyes enlarged and unblinking, his heart and lungs racing. As the pouring rain swept inside the topless barn, it blew away from him, soaking the opposite wall.

Relatively sheltered once more, Willard's vital signs returned to normal. In only fifteen minutes, the rage of the storm, too, began to abate. Willard stood to avoid being soaked through as the rainwater rolled nearer across the barn floor. Still flat against the wall, though, he began scanning the tops of the timber posts from which the roof had torn free. Something about the remnants seemed odd, unnatural.

Soon, as the back side of the thunderheads receded toward the east, the rain subsided to intermittent drops. With the thunder still echoing in the distance, Willard set a ladder against the barn wall and climbed to inspect the tops of the posts. The break was clean. The roof had not been torn from the structure, nor peeled like the lid from a tin can of tuna. It had been removed, lifted straight up, like his daughter Vivian did with her doll house.

How, he wondered, could he account for that? He found the roof intact some distance away. Understanding how it got there would help him understand how to repair it, but it would also answer a new question: How does a roof fly?

Willard had been enthralled with the dynamics of flight since he could remember. Wilbur and Orville Wright had made their inaugural flight in 1903 when Willard was four years old. These men lit a flame that engulfed the world. By 1908 the Wright brothers had honed their plane's controls to the point that they could stay aloft for an hour—banking, climbing, and then landing with grace. But competition had already grown keen. In 1908 in Hammondsport, New York, Glenn Curtiss made his first public flight in his plane, June Bug. A motorcycle engine mechanic with an eighth-grade education, Curtiss refocused his mechanical expertise to aircraft and soon claimed fame with rising aerial speed and distance records. Also in 1908, Henri Farman in Europe completed his first flight.[4] Within six years he founded a construction company that joined others to invent and produce Europe's first fleets capable of waging war from the sky.

Throughout World War I—Willard's teen years—newspapers described swarms of allied and enemy planes carrying bombs, machine guns, and reconnaissance maps above the battlefronts. With the end of the war and the beginning of the 1920s dawned the Golden Age of Flight. Daily, the newspapers trumpeted the names of heroes with newly designed aircraft, prize winners of new flight records, and other aerial marvels. New pilots, home from the war with surplus biplanes, barnstormed the countryside like dragonflies, buzzed rural towns, performed aerial acrobatics, and offered local folk inexpensive tours of their landscape from the training seat of a biplane. An equal opportunity employer, barnstorming allowed men, as well as women and minorities, to earn a few dollars, and introduced names like Charles Lindbergh, Bessie Coleman, Clyde Pangborn, Wiley Post, and Harriet Quimby to the public. The western world spun in the whirlwind of powered flight.

Willard was a product of this age. Like Glenn Curtiss, he was a mechanic with little more than an eighth-grade education. But he devoured all he could forage about airplanes and how they worked from his older brother Cecil's school textbooks, the local newspapers, and the magazines he found at the newsstand on every errand that took him into town. Unlike Curtiss, he never became a pilot, but the mechanics of how an airship, bearing hundreds of pounds, could float on a breeze never ceased to intrigue him.

So, this event, in which a storm had decapitated his place of refuge,

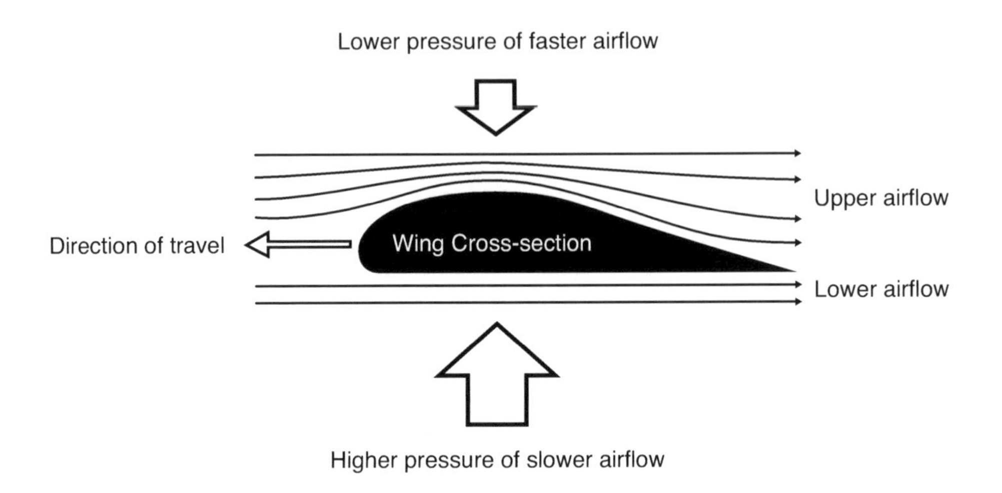

Figure 1.1 — Components of lift around the airfoil

presented Willard with a mechanical airplane puzzle. While being familiar with the basics of flight dynamics, he could not fit together all the pieces: How could a barn roof, which is not designed for flight, achieve what aircraft to date had not achieved, namely, to rise from a standstill and simply fly away?

The Idea

Willard knew that an airplane flies because the stream of air above the wing travels faster than the stream of air below the wing. The difference in speed between the airstreams creates a corresponding difference in air pressure. The pressure below the wing becomes greater than the pressure above the wing, and so it pushes the wing upward. This is called lift (Figure 1.1).

Knowing at least these basic facts about lift, Willard realized that similar forces had acted on the barn roof. Because of the storm's ferocious wind speeds, the air above the barn roof had been traveling much faster than the practically stationary air inside the barn. The fast airflow had consequently lowered the air pressure above the roof. He reasoned, then, that the higher air pressure under the roof (inside the barn) must have forced the roof up, tearing it from the timber posts and serving it to the storm winds. Put simply, he believed the roof had literally popped off the barn walls (Figure 1.2).

Over the weeks and months that followed, Willard pondered the implications of this for an airplane. He visited the library in Hagerstown, researching the books and journals devoted to aeronautics. Every story

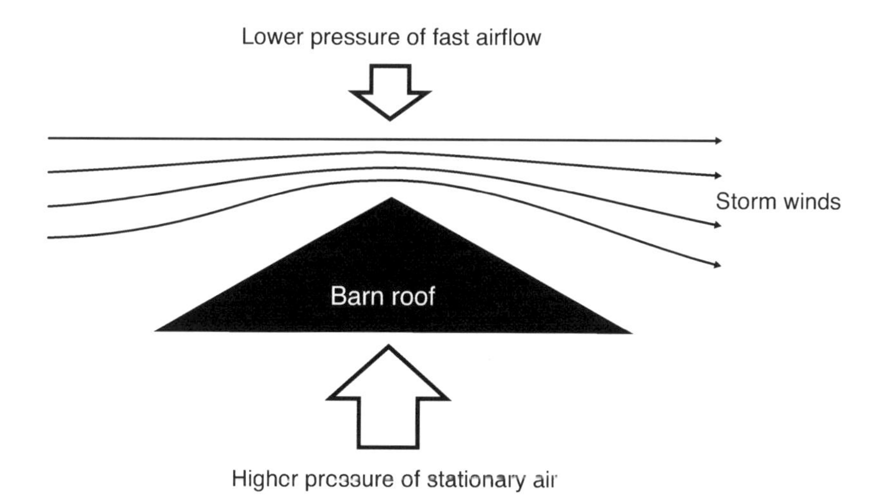

Figure 1.2 —Components of lift around the barn roof

involved gliders being catapulted or dropped into the air, or motorized airplanes with wheels or runners or pontoons, whisked across land, ice, or lakes to get them into the air.

In the conventional aircraft of his day, an airplane wing would move through the air at a fast speed (called airspeed) to generate lift for takeoff. However, because of his recent experience in the storm, Willard came to understand that air rushing at takeoff speed (the speed of air, he called it) over a stationary airplane wing would also produce lift. That meant that a stationary airplane (one with no airspeed) could take off as soon as the air coursing over its wing (the speed of air) reached takeoff speed.

If this was true, could he create and control such a flow of air?

His First Patent

In 1925, Willard lived with his wife Lula and three young children (Vivian, Harold nicknamed "Curley," and Helen) on the west end of Hagerstown, on property he had bought the year before.[5] He would live there with Lula the rest of his life and add to it over the years to make it a haven. With the house, garage, Concord grape arbor, apple tree, chicken pens, small outbuildings, garden, and workshop, he had made it a model of self-sufficiency.

Lula was petite and the classic acquiescent wife. Behind her wire-rimmed glasses, she had the benign round face of a tabby cat. She wore her mouse brown hair braided and wound around the crown of her head. At home she always wore a cotton housedress, knee-high hosiery, and black leather, chunky heel shoes—always—even when gardening or feeding the chickens, except that then she added galoshes and a wide-brimmed straw hat. In the kitchen, she simply added a bib apron.

One Sunday after washing lunch dishes, Lula found Willard on the back porch with a 6.5-ounce Coca-Cola bottle, a pencil, and a pad of yellow, unlined paper. The children were playing nearby in the yard.

"What are you working on?" she asked. "Do you still have that storm on your mind?"

He smiled to say Yes.

"A storm on your mind...I'd call that a brainstorm."

He grinned at her word play.

"I'm working on one," he acknowledged. She knew him well, he thought, and that comforted and encouraged him.

"I'm going upstairs for a nap, so I'll be fresh for bridge tonight," she said, handing off childcare duties without saying so.

"Okay, Mother. Sleep well," he said, returning to his sketch.[6]

His first idea was to use a spinning propeller to generate the airflow over the wing. The purpose of a propeller is to move an airplane forward, but there's no missing the fact that, like a fan, it generates quite a breeze in the process. Capitalizing on that breeze, Willard imagined he could place a wing directly in the propeller's wash and thereby bring the air to the wing rather than taking the wing through the air. If the speed of the air were sufficient, the result should be enough lift to raise an aircraft.

For his first solution, as documented in his first patent, he began with a standard single-engine biplane, the conventional airplane of the 1920s. This consisted of a fuselage sandwiched at the front between two wings. On the nose, a single engine turned a propeller. To create lift without moving the airplane down a runway, Willard proposed to simply add a pair of winglets—"lifting members," he called them—along each side of the fuselage directly behind the propeller. Placed here, they would be in the airflow blowing from the propeller and washing over the length of the fuselage, thereby creating static, vertical lift.

The problem with using a propeller in this way, however, is that the purpose of a propeller is to move the plane forward—horizontally. But Willard did not want the plane to move forward. He wanted it to hover like a hummingbird even after it left the ground.

He knew that the propeller, as it turns, screws into the air mass in front

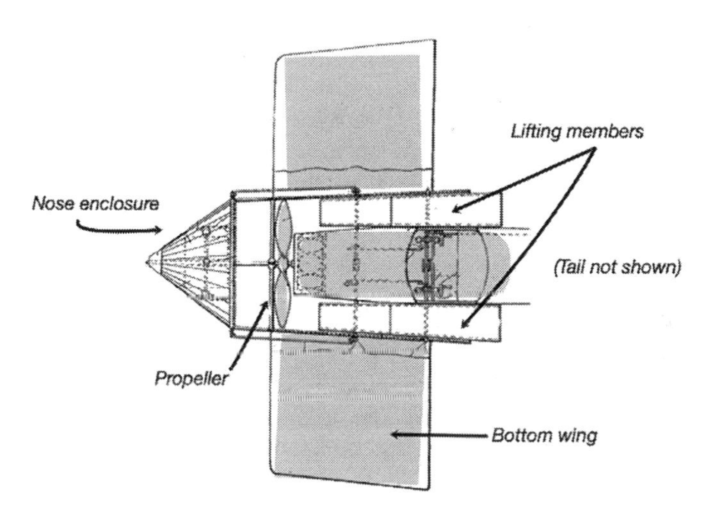

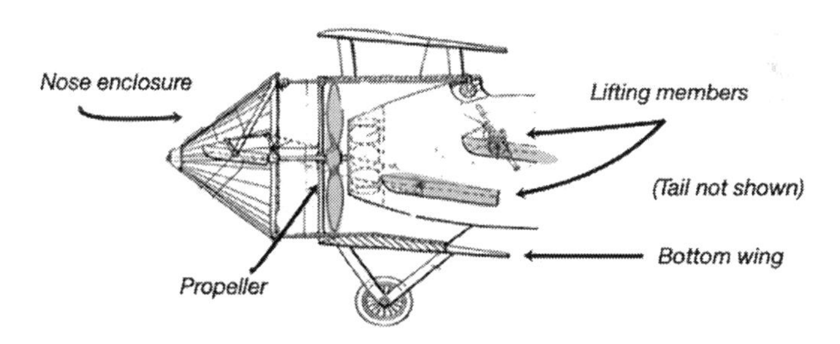

Figure 1.3 — Diagrams from Willard's first patent (shading added)

of it, thereby pulling the plane forward. So, he wondered, if he walled off the propeller from the air in front of it, like closing the window in front of a window fan, could he cancel the forward motion? But wouldn't that also reduce the breeze it generated? Maybe he could supply air to the propeller from a different direction.

He proposed, then, to encase the propeller—and, with it, the entire nose of the plane—in a cone that would control the air around the propeller. The frame of the cone would support a series of overlapping slats that, like Venetian blinds, could be opened and closed. When they were open, air could flow directly to the front of the propeller, and the plane could move forward. But when the slats were closed, air could enter the enclosure but only from an opening in the top. In this way, he thought, he could start and stop forward motion without reducing the airflow to the "lifting members" (Figure 1.3).[7]

Two aspects of this design are important. First, it shows one way to place wings directly in the wash of the propeller to generate lift before the plane starts moving. Second, it incorporates a means to brake forward movement—even in the air. That's right: air brakes.

At least, that was the idea.

As step one in a larger strategy, Willard chose to patent this idea and applied to the U.S. Patent and Trademark Office (USPTO) for a design patent on December 22, 1927. Patenting his design would protect his right to exclusively sell his idea in the marketplace, should that opportunity ever materialize. While a design patent protected only the appearance of an invention—not the details of how it worked—Willard could more easily get a design patent than the conventional "utility" patent. It would still protect Willard for several years if he chose to further develop his idea.[8] The patent attorney's bill and the patent application fee discouraged frivolous applications but were not unaffordable for Willard. This patent was granted on April 9, 1929, a year and a half after he applied.

His First Experiments

While awaiting his patent, Willard experimented with various components of the design in his workshop.

He had built his workshop where the backyard met the untilled fields fenced off for grazing or cultivation. Lula's large garden plot isolated it from the county road. Willard usually walked to "the shop" from the house via the narrow gravel path that navigated between Lula's grape arbor and the detached garage, following it through the backyard, alongside Lula's prized peony bushes and her lush lilacs. The path then skirted the limestone erupting like a scab from the ground on the left, and passed the pens of chickens, every bird named by Lula. The path ended at the two wooden steps at the shop door. Leaving her world, he entered his.

The shop itself consisted of a simple white clapboard building of two levels, "a chicken-house laboratory"[9] set into a slight grade. On the front at the near end, the door to the workshop opened onto the upper level. At the far end the two-car garage emerged from below ground.

The workshop filled half the upper level, an open room that felt too small. It housed a lathe, table saw, band saw, drill press, and two tables with mounted vises and overhanging light. It perpetually smelled of engine oil, wood shavings, and sawdust, especially that of the aromatic cedar which grew in scrubby patches in the adjacent fields. Between projects Willard would have kept it swept and tidy, but his projects always overlapped. He mounted the tools of the mechanic and carpenter trades, many of which he had made by hand, like deer heads on the walls.[10]

The workshop had become Willard's favorite place, a tangible expression of his inner self. Here he poured his efforts and creativity into one project on top of another. And here he proceeded to experiment with controlling the flow of air in and around a propeller.

As an engine mechanic who worked amid discarded engine parts, he could easily build a simple engine from scraps. For one of his experiments, he anchored such an engine to the shop floor and mounted a propeller onto its crankshaft. He then built a three-sided box that enclosed the engine and propeller on the bottom and the two sides, leaving the top, front and back open. Thus, the propeller was housed in a trough.

How would the air behave around a propeller rotating inside a kind of box?

Willard switched on the engine. It thundered inside the workroom, turning the propeller blades so fast that they disappeared. He let the engine warm a moment while his ears adjusted to the decibels, then gradually he increased the rotations per minute, his hand poised over the shutoff switch. He noticed nothing remarkable, and he pushed the throttle further. Then, further. Then, too far! With a crash, the workshop windows imploded. The sides of the trough collapsed inward, funneling broken glass and boards into the sucking vortex of whirring propeller blades.

Instantly shutting off the engine, Willard froze. He let the shock drain from his body as stillness returned and the debris cloud settled. He was rattled but physically unscathed. The inside of the shop, however, looked like a tornado had hit it.

Cautiously, almost apologetically, Willard tiptoed outside to get some fresh air and to clear his thoughts. He was developing a healthy respect for the powerful, invisible forces of the atmosphere in which he was submerged. As he wiped sawdust from his glasses, his mind explored the implications of this latest experience. By funneling the airflow through a trough, he had temporarily created a "vacuum," as he understood it, a low-pressure pocket of fast-moving air immediately in front of the propeller. The greater air pressure surrounding the trough and outside the shop—approximately one ton per square foot at sea level—had reacted irresistibly to fill the vacuum, almost to the point of crushing walls. Luckily, he had left the shop door open, he consoled himself, or the damage to the shop might have been much worse.[11]

If this explanation was accurate, he imagined, he could change his airplane design to incorporate a similar trough. A low-pressure trough attached to a stationary aircraft could make it extremely buoyant. A plane so designed might even cut a vertical flight path to the sky.

A Working Model

Willard's successes with these part-time experiments pumped adrenalin into his blood. His experimentation had become an avocation, and he needed more money to continue. The son of a blacksmith, and an automobile mechanic with no formal education beyond his thirteenth birthday, he had little money of his own. He had a job, but he also had a mortgage, and his wife and children to feed and clothe. Worse, the Great Depression was dismantling the economy around him.

But he found an opportunity. In the 1930s, he moved from fixing cars to selling them. With his affability and boyish enthusiasm, he proved to be a salesman who could close deals. Even during the 1930s he averaged more than $100 per week when that salary seemed like a fortune. He learned how to make money flow, and then to make it flow his direction. For example, if a potential buyer approached him for a make and model of automobile that couldn't be found on the car lot, Willard would note the information and promise to call if the car became available. Then he would scan the traffic as he commuted over the next days and weeks. If he spotted a vehicle matching the description, he would follow its driver to his destination and then ask him if he'd be willing to sell it. If so, Willard would broker the deal and take a commission.[12] In this way, he transitioned from the garage into the car lot and the public stream.

As another child (Kenneth nicknamed "Reed") was born and the children grew, Bluegrass music filled the house in the weekend evenings, with Willard playing banjo, the daughters playing piano (one piano, two girls: one girl playing the left hand, the other playing the right), and two sons picking up fiddle and guitar. Lula was their initial audience, and the children would eventually go on to win local talent contests. Many years later, Curley would be famous on the Atlantic seaboard as a square dance caller, and Reed would be a professional bass musician.

Leveraging this interest, Willard started to sell musical instruments. In 1937, he started his first company, the Federal Violin Bureau, which for the next four years employed a lone French violinmaker to make and repair stringed instruments.[13] Willard found it surprisingly easy to start a company as it involved little paperwork. The company rented space over the Colonial Theatre in downtown Hagerstown, and Willard spent more time in town.

Several of the floors above the violin maker were vacant at the time, and since Willard stayed in town and away from his workshop, he temporarily used an upstairs room in the theatre to build and test his first airplane model (Figure 1.4). His oldest son, Curley, now in high school, stopped by after school to help with it.

The model was constructed primarily of a translucent material stretched over a wood and wire frame. It consisted of a fuselage with two short wings that ended at uprights. The total wingspan measured nearly five feet. Behind each wing, a small ¼-horsepower gasoline engine was attached to the fuselage. Each engine held a propeller positioned behind the respective wing but in front of the engine, that is, in "tractor" position.[14]

In the wings were "channels," as Willard called them—square troughs

Figure 1.4 — First aircraft model

through which the propellers would suck the air at high speed. A long, flexible gas line ran from a small gas can to each motor. Willard opened the windows for ventilation, taking every precaution for safety.

To test the model, Willard placed it on a table and tied a string to the model's tail. He started the engines, and the spinning propellers instantly became blurs, the pungent odor of spent fuel permeating the air. Willard pinched the string between his thumb and finger—the model plane was trying to crawl forward across the table. He then increased the engine speed. As if by magic, the craft rose straight up from the tabletop. Had it not been tethered, the plane would have flown up and collided with the ceiling. Instead, it hovered for an instant as high as Willard would allow, and then he slowly cut the fuel supply. The model gently dropped onto the tabletop.

The model worked so well that Willard drafted an application for a utility patent based on it. In it he described his model's features and briefly explained, mechanically, how they worked.

A Quaint Corporation

A gregarious sort, energetic and fast-moving, Willard quickly met other business owners and members of the Hagerstown Chamber of Commerce. One of them was Frank Kelley (Figure 1.5).[15] Entering Kelley's Studio and Camera Shop just two blocks from the Colonial Theater, Willard instantly liked Frank. He had thick black hair, a long, pleasant face and nose, and dark eyes. Frank had grown up an Irish Catholic in the Midwest, and he embodied all the best attributes of that environment. Born and raised in Canton, Ohio, his warmth, intelligence, and principled ways gave him much in common with Willard. He had embraced photography in high school, experimenting with camera equipment and film-developing chemicals in his spare time. He'd also earned his private pilot's license and bought his own plane, combining his passions into aerial photography. Recently, he

Figure 1.5 — Early Photograph of Frank Kelley (1909-1996)

had come to Hagerstown to set up shop. Entrepreneurial and savvy, he was also president and treasurer of Tri-State Tank Lines out of Baltimore, a firm operating petroleum tanker trucks.[16]

Willard engaged him to photograph his model. But when Frank saw Willard demonstrate it, he responded, "Willard, we need to talk."

The next morning found them in a downtown restaurant, huddled over coffees and a notepad. Frank outlined for Willard all the advantages of forming a C-corporation in contrast to the sole proprietorship he had with his violin company. They talked about a business plan, a bank account, shares, a board of directors, officers, registering with the state of Maryland, and next steps. Willard trusted Frank. They ended the conversation with a handshake, a smile, and a list of tasks each would be responsible for. When they left the café after two hours, they walked out excited, with plans to meet again in a few days (Figure 1.6).[17]

As a result, Willard started his second company in 1939, the National Aircraft Corporation, devoted to developing, protecting, and marketing his aeronautical ideas. He also quit his full-time automotive sales job.

He came to that decision easily but with trepidation. Considering all he had learned, and all he had invested in time and money, committing to his company fulltime felt natural. But he didn't feel good about certain opportunity costs, such as the alienation of some extended family members and friends. These now avoided him, tiring quickly of his repeated suggestions that they "get in on the ground floor" and buy stock in his company. Others berated him, calling him irresponsible to risk his family's income during an economic depression, and on a dubious airplane idea. Some called him crazy and shook their heads, laughing at him as they quickly walked away. But Lula stuck by him. If she ever second-guessed him, she never voiced it.

His corporation was quaint, what *Flying* magazine called "a symbol of small-town neighborliness and enterprise." Willard served as chairman of the board with Frank Kelley as president, the only board member who could fly a plane. The other board members included a grocer, a baker, a dealer in wholesale fabrics and trims, and a lawyer practicing in Washington after serving three years in the Navy.

The company's corporate office was a tiny room in Willard's backyard workshop, the corporate phone number matching Willard's home phone. Willard kept the workshop locked in his absence and used a rural intruder alarm system: behind the workshop, a large pen, chicken-wired over the top, housed and barely restrained two or three families of ring-necked pheas-

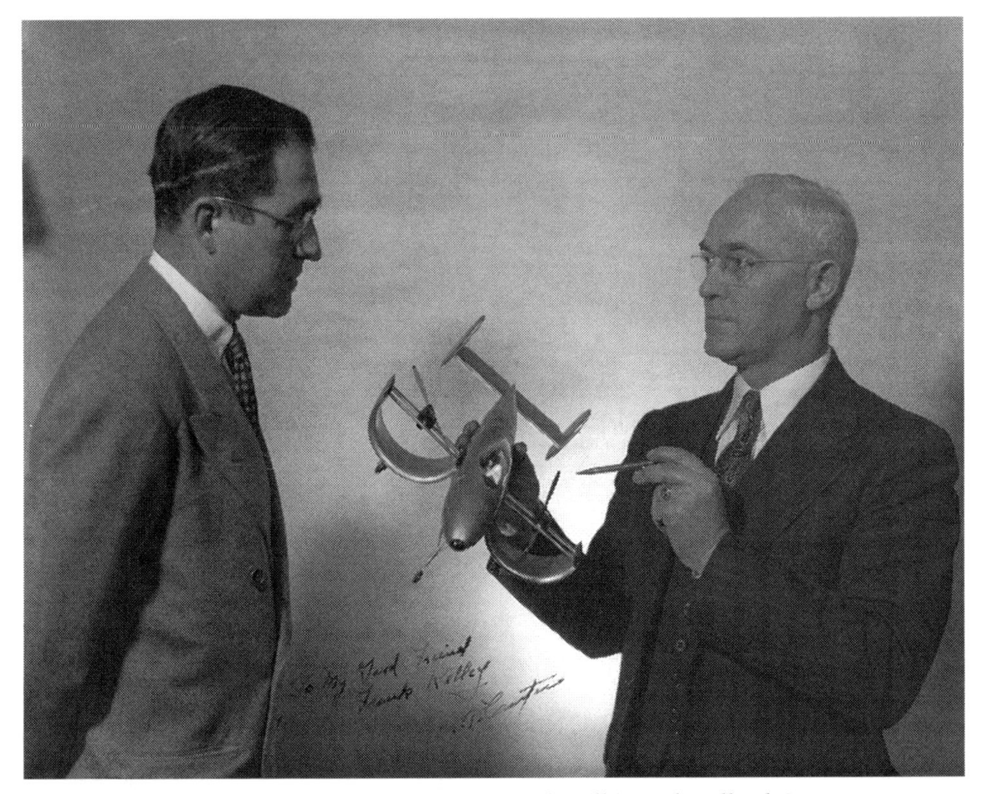

Figure 1.6 — Business partners, Frank Kelley and Willard Custer

ants. Willard and Lula enjoyed eating their little brown eggs. The untamed hens accepted them, allowing them to retrieve the nuggets without alarm. But if the pheasants caught sight of anyone else, they crowed in terror, launching themselves dangerously into the wired walls of the pen, frantically fighting to flee. In so doing, they raised an alarm that immediately registered at the house. This alarm system remained continuously active and required no electricity.

Willard would soon be glad to have it.

Fitted out with a corporation and a working model early in 1940, Willard approached Bernard Garvey, Esquire, an experienced Washington, D.C. patent attorney. At fifty years of age, Garvey had been practicing patent law in the nation's capital for twenty-five years. A graduate of Georgetown University Law School, he had also been an adjunct professor of patent and corporate law at Columbus University (Washington, D.C.) for ten years. Soon, beginning in 1945, he would be an adjunct professor of patent law at Georgetown. His law offices were enshrined in marble and brass at 1010 Vermont Avenue NW, just three blocks from the White House. Willard retained Garvey to file his patent application as the next step toward industry success.[18]

This application, entitled "Aircraft Having High-Lift Wing Channels," differed from his first one in that this application was for a utility patent, so-called because it must describe an invention that actually operates. It described his model, and the model clearly worked, so Willard anticipated the Patent Office would grant it easily. Willard employed elementary concepts from Curley's physics textbook to describe his model and Garvey filed the application August 31.

In a brief discussion in Garvey's plush office, Willard explained the genius of his concept, and he gushed about his intent to eventually build and fly a full-size airplane of similar design.

Sizing up his new client, Garvey asked, "How much have you spent so far on your experiments and your model?"

"Four or five thousand dollars," Willard confided, estimating his expenditures over the past fifteen years since the barn roof disaster.

"Four or five thousand? Humph! Well! I don't know how familiar you are with the money involved in making an airplane fly starting out from scratch," Garvey started to say, frowning. But then he assumed a more fatherly tone and continued, "It is not unusual, according to my information, to spend from four to five million dollars, and when you spend from forty to fifty thousand dollars in experimentation, it is only a drop in the bucket."

Garvey let Willard measure his meaning. Willard sat back into the tight leather chair and exhaled audibly, unable to comprehend a million dollars. But his enthusiasm continued to churn behind his caramel eyes while he re-assessed his circumstances. Willard had no hope of making any appreciable progress against the costs Garvey was describing with the nickels and dimes he'd pooled by personally approaching friends, family, and acquaintances. He also realized Garvey was savvy about patents and the airplane industry, and he needed this man's help.

"What do I need to do?" Willard asked.

Garvey suggested that Willard needed what today would be called a corporate venture capitalist with some significant seed money, a company that would be willing to take a risk on Willard's idea and provide the funds required to develop it. The car companies were heavily investing in the aircraft industry. Maybe one of them would be open to helping him, Garvey suggested.[19]

Willard left Garvey's office on a quest. A few months later, he hopped on a train to Detroit, the next stop on a career he could only begin to imagine.

II The Bumblebee 1940 - 1943

When they heard about me, they knew I was crazy, they knew it wouldn't work. And I didn't know it wouldn't work. It was like the [B]umblebee. I didn't know it wouldn't fly, and I went ahead and flew it anyhow. But the smart boys who had all the education knew it couldn't work, so I didn't have to forget that.[1]

- Willard Custer

In the autumn of 1940, 41-year-old Willard stepped into the reception area of the Briggs Manufacturing Company's corporate office in Detroit.[2] The owner, millionaire industrialist Walter O. Briggs, also owned the Detroit Tigers baseball team. He had grown his company into the largest car manufacturing company in the country, producing car tops and bodies for Ford, Chrysler, Chalmers, Hudson, and Packard, all at the same time. Within a few years it would employ forty thousand workers at sixteen plants, nine of them in Detroit.[3] Willard knew the name Briggs because he knew every part of a Ford automobile. With most of the car manufacturers in the country invested in the aircraft industry, Willard went straight to the top of the list.

He wore his best black pinstripe, three-piece suit with his favorite tie and fedora. With his black shoes spit-polished, he felt confident. Standing five feet and seven inches tall, weighing a muscular 180 pounds,[4] with caramel brown eyes and bushy black eyebrows that contrasted prominently with his already graying hair, he looked the part of the Chairman of the Board of the National Aircraft Corporation. In his briefcase he carried a copy of the patent application he had filed in August and a reel of film.

Willard knew that before Briggs would put any money into his channel-in-wing idea, Briggs would want to know that Willard had control of the design in the marketplace via a patent.

It was not Mr. Briggs, but Howard Bonbright, the company treasurer, who greeted him and invited him into his executive office. He introduced Willard to three of his assistants, exchanged chitchat about Willard's trip from Hagerstown, and then invited him to a table prepared with a projector and screen so Willard could show them his idea.

Seated, Willard related his experience in the storm several years before. Then, in his choppy, West Virginia accent, he described the principles of flight, especially as they related to his barn-and-gale experience. And then he explained how the "speed of the air" was at the crux of airplane lift; and how, if it could be channeled across a specially designed wing, it would cause an airplane to rise vertically—yes, even from a standstill.

Willard then presented the silent 16mm film of his model that Frank Kelley had recorded. While the projector clattered, Willard directed his audience's attention to the wings, forming what he called "channels," square troughs through which the propellers would suck the air at high speed. Low-pressure air pockets would form in the troughs, he explained, and the craft would rise vertically from the table.

Willard also pointed out the string tied to the model's tail. "I don't want it to fly into the ceiling," he explained. "It's the best model I have, and I have several thousand dollars invested in it."

The men watched stoically as the engines started and the model jumped to life. Then they watched the craft rise straight up from the tabletop, and none but Bonbright could hide his blink of astonishment.

Willard stopped the projector.

In the returning silence, Bonbright responded in a low voice. "I'm very interested," he said, almost gravely. "Do you have a patent for this?"

Willard handed him the application. "It isn't approved yet, but it will be."

"May I keep this?" Bonbright asked, handing it to one of his assistants.

He proceeded to interview Willard, learning more of his educational background, sizing him up against his own Ivy League scale. He learned that the closest Willard had come to an engineering education of any kind had come by way of his father. Clem Custer had been a village blacksmith in rural Pennsylvania, Willard explained, but he had taken a locomotion mechanic's job in the Baltimore and Ohio Railroad's rail yard at Martinsburg, West Virginia. Many blacksmiths like him had become the first mechanics in the burgeoning transportation industry. Clem had trained Willard to fashion whatever he needed with his own hands and to understand how things worked from the mechanical standpoint.

Willard's formal schooling had ended just after he turned thirteen years old, he told Bonbright. His older brother Cecil had finished high school, and then attended college and law school while working for the Civil Service Commission in Washington, D.C. But economics had forced Willard home to help his father meet the needs of the household.

Bonbright could see that Willard felt neither ashamed nor proud of his circumstances. They were mere facts.

"You live near Washington?"

To Willard's nod, Bonbright turned to an assistant.

"Do we have someone in Washington?"

"Yes, sir," came the reply, and the man handed him a sheet of paper. Bonbright read it, then handed it to Willard, who saw that it contained a man's contact information.

"Mr. Custer, here's what I'm going to do," Bonbright concluded. "I want to help you, but I don't know airplanes; I know automobiles. So, when you get home, you will contact this man, Reuben Davis. That's his information. He's a consultant with the Curtis Wright Aeroplane Company in Washington. I will call him today and tell him to expect your call."

"Yes, sir?" Willard paused for more.

"If he tells me he is satisfied that your idea has merit," Bonbright continued, "I'll draw up the agreement and give you enough money to build it."

Willard wanted to cheer and groan at the same time. Bonbright had made quite a promise, but maybe Reuben Davis operated as a mercenary, hired to kill ideas.

"Do you have any questions?" he asked, standing, shaking Willard's hand.

Willard had a hundred of them, but Bonbright was walking toward the door.

"Thank you very much for your time," Willard replied, gathering his film.

"Let me know when that patent gets approved," Bonbright said, as Willard walked out the door.[5]

First Channel Wing Scientist

Soon after returning to Hagerstown from Detroit, Willard contacted Reuben Davis, as Bonbright had directed. Davis instructed him in turn to send the film of his model to Dr. Louis Crook in Cabin John, Maryland.

Dr. Crook was then the Dean of the Department of Aeronautical Engineering at The Catholic University in Washington, D.C. He and his mentor, Alfred Zahm, Ph.D., had built the department from the ground up. Dr. Zahm, formerly professor of mathematics and mechanics at Notre Dame University, had come to teach at The Catholic University in 1895. Six years later, and two years before the Wright brothers flew at Kitty Hawk, he had provided the specifications and overseen the construction of the college's aeronautical laboratory. It was a small affair, 30 feet by 80 feet and one story high, but it housed the first significant wind tunnel to be built on any college campus in the United States. The wind tunnel dominated the inside of the building, with height and width measuring 6 feet, and 40 feet in length. A 12-horsepower electric fan generated the "wind."

In addition, Dr. Zahm had invented instruments for his wind tunnel to enable precise measurement of air velocity and air pressure, and how each varied with different objects in the air stream. In 1902, Dr. Zahm presented a paper describing his laboratory and detailing his work to the American Association of the Advancement of Science. His work on airplane hulls and fuselage shapes in 1907 led to the conclusion that the blunt torpedo shape (thereafter called the "Zahm shape") encountered less air resistance than any other shape, and the young airplane industry soon adopted it as the standard fuselage design.

Dr. Zahm had integrated his wind tunnel experiments into his classroom, involving his students in his research. He felt particularly drawn to one of his students, five-foot tall Louis Crook, the bright, hard-working son of the university's former general manager. Dr. Zahm left the university the year before Crook graduated, but he had forged a relationship with the young man that would last the duration of their professional lives.

Crook later graduated and entered a doctoral program at Johns Hopkins University as his mentor Dr. Zahm had done. He then assumed Dr. Zahm's wind tunnel studies as part of his own graduate work. At this time, the tireless young man also taught physics as an assistant professor at Catholic University, lectured at the nation's Bureau of Standards, and worked as a machinist at the Navy Yard from late afternoon until midnight.

In 1916, Crook earned his doctorate in mathematics and physics from Johns Hopkins University and then returned home to Catholic University as a full professor in theoretical and applied mechanics. Meanwhile, Dr. Zahm was named Director of the Aerodynamical Laboratory of the U.S. Navy, and he soon enlisted his former protégé in his work, part-time, testing airplanes in wind tunnels for the Navy. The two wind tunnel pioneers co-authored the first textbook on airplane stress analysis.

Dr. Crook had risen quickly to become the university's Director of the Department of Mechanics. Then in 1935, Catholic University established its Department of Aeronautical Engineering and named the formidable professor its Dean.

Dr. Crook continued his wind tunnel development and research. Over his tenure he built another wind tunnel for the university, wrote 337 wind tunnel reports, and pursued interests in guided missiles and supersonic aircraft. He also built several wind tunnels on his extensive private property in Cabin John, Maryland. He used these to conduct research for companies and individual inventors on a contract basis.[6] In the autumn of 1940, Reuben Davis sent him Willard's film.

"I want to make some preliminary tests before I make any suggestion," he responded after viewing the film. At his request, Davis and Willard delivered the patent diagrams and some models to him at his residence in Cabin John. Davis introduced the two men.

"Doctor Crook," said Willard, shaking his hand warmly. The dean stood shorter than Willard.[7] "It's a pleasure, Doctor. I'll trust that your last name does not describe your character," he dared to venture with a boyish grin.

"Thank you, Mr. Custer," responded the professor with a straight face. "I'm sure many people have thought to make that joke, but you're the first one to say it to my face."

Pausing just long enough to let Willard wonder if he had offended, Crook then smiled, and quipped in return, "And I'll trust that your last name is no omen of your prospects for success," a reference to General Custer's demise in the Little Big Horn battle.

"Why, thank you, sir," Willard chuckled. "I hope to come out better."[8]

One day soon afterward, the mailman delivered a letter from the Patent Office to Willard's mailbox. It contained the response to his application for the utility patent. "This device won't work," it said. "It is inoperative."

The patent was denied.

On reading it, Willard stood momentarily stunned. Garvey had explained to him that relatively few patent applications are issued as originally filed, but Willard did not expect an instant denial. This patent was critical. Without it, everything that had gone before became a waste of time, effort, and money. Without it, he became an embarrassment to his family and in desperate need of a job. He could not accept "Denied." So, he quickly dialed Garvey, his patent attorney, who had just received a copy of the same letter.

Garvey explained that, despite the definitive wording, the denial was not final. In responding to a rejection notice, he must address each of the patent examiner's objections and request reconsideration. Statistically, patents were ultimately granted in two out of every three applications, Garvey reassured him.

So Garvey responded to the patent examiner's objections within the allotted six months, and then he initiated a series of meetings to discuss them in person.

Over the winter months, 1940-41, Willard drove the 70-mile trek to Cabin John, Maryland, multiple times a week, meeting with Dr. Crook as he tested Willard's model in his private wind tunnels (Figure 2.1).[9] He also traveled even farther to Washington, D.C., several times, to meet with the Patent Office examiners.

Willard loved explaining his invention to anyone with a curious mind, but he quickly lost patience with the Patent Office examiners. Their minds

Figure 2.1 — Dr.Crook experimenting on a model Channel Wing

were closed behind deadbolts, it seemed, and the more he met with them, the more the meetings turned into arguments. The examiners were well versed in the conventional aerodynamic theories of the day, being college graduates. Willard, however, had no such degree, and the examiners knew it and discounted him because of it. But Willard had eyewitness evidence, and these scholars could not stack their arguments high enough to convince him to distrust his eyes. Willard concluded, therefore, that the education the examiners had gained had prejudiced them, had disabled their ability to think in new ways, and he felt fortunate to be without it. "The smart boys who had all the education" became his derogatory way of referring to them.

Tensions soon led Garvey to meet with the examiners without Willard (his office sat only six blocks away), but those meetings would be less than helpful. Garvey could not adequately explain the workings of Willard's invention, and he sometimes returned from the meeting with more questions than answers.

Fortunately, Willard had continuing access to someone who could help him: Dr. Crook. Willard took these discussions to Dr. Crook, an expert in aeronautical engineering. Dr. Crook taught him some engineering terms and concepts he could use to rebut the examiners, in the process providing Willard with the aeronautical engineering education tailored to his aircraft.

Patent protocol insisted that Willard's explanations had to be put on paper. The patent examiners could entertain all the conversations they cared to, but, by the rules, their decision must be based on documentation alone.[10]

"Since I find it almost impossible for someone other than myself to explain the theory and operation of this new type airship, I wish to supply a resume that can be available for reference when I am not present," Willard wrote Garvey on June 18, 1941.

In the document, Willard described what happens to the air currents around a stationary propeller. He explained that, as the spinning propeller pushes the air behind itself, an area of reduced air pressure forms in front of it. Some of the air that the propeller has pushed behind then circulates outward, around the propeller tips, and back to the front of the propeller, drawn there by the area of reduced pressure. Willard then pointed out that if the propeller is positioned very close to the back of the channel, then the channel obstructs this passage of air back to the front of the propeller. This obstruction further reduces the air pressure within the channel, thus creating even greater lift potential.

Willard continued:

> At 1500 RPMs the wing section has ceased to weigh anything

> as the atmospheric pressure under the wing exceeds the weight of the wing section and the wing flies for the same reason a roof leaves a building during a storm...Now we have a low pressure area over a wing section of tornado proportion that can be increased or lessened by the RPMs of the propeller.[11]

Willard signed his letter, "W.R. Custer, Engineer." His role in the project and five months' education working with Dr. Crook—an engineering immersion course—had turned him into an engineer in his own right.

In the end, however, neither the conversations nor the documentation would convince the examiners.

Finally, Garvey got permission from the principal examiner to present the film of the model demonstration, even though it would not qualify as documentation. He would later report that the film was shown "in the presence of the Principal and Assistant Examiners—in fact, nearly all the Examiners of the Division were present when this demonstration was put on...As soon as we completed the exhibition of the picture, the Principal Examiner... said, 'We are satisfied...that the device will operate.'" The examiner immediately indicated his willingness to sit down with Garvey and work on the wording of the patent with the intent to issue it.

This little piece of film was proving to be priceless.

About this time, Frank Kelley came across a one-page magazine ad for the Bridgeport Aluminum Company that made him smile. The ad stated categorically, in tones reminiscent of the patent engineers, that "The Bumblebee Cannot Fly!" It went on to explain, "According to recognized aero-technical tests, the bumblebee cannot fly because of the shape and weight of his body in relation to the total wing area. BUT, the bumblebee doesn't know this, so he goes ahead and flies anyway!" Frank tore it out, fed it into his typewriter, and typed under it, 'CHANNEL WING DOES THE SAME,' and gave it to Willard.[12]

On December 7, 1941, the attack on Pearl Harbor spurred the United States to enter World War II. Nineteen days later, after a year performing his preliminary work with Willard's models, Dr. Crook wrote the following report, "on the result of experimental tests made to determine the adaptability of the Custer airplane principles to a full-scale aircraft":

> From the experiments made by me, I am satisfied that when the Custer principles are applied with scientific accuracy for a specific use, such as accelerated lift, low landing speeds, or high speed flight conditions, substantial improvements will be found.
>
> My experiments show that the Custer propeller-wing combination can be so designed as to substantially improve both take-off and landing so that the engine horsepower can be used to its greatest advantage. Further, a light ship can be more readily controlled during gusts or other adverse weather conditions.
>
> The general conclusion from all the tests made is that the principles found in the Custer airplane are sound and practical.[13]

Dr. Crook also recommended another series of wind tunnel tests that he proposed to conduct. The report convinced Reuben Davis, and he forwarded his recommendation to Bonbright. Soon after, Willard and his business partner Frank Kelley put Dr. Crook on contract to continue his testing, with the expenses to be covered by Briggs Manufacturing. In all, Dr. Crook would conduct literally hundreds of tests on Willard's models, as well as similar models of his own making.

The First Channel Wing (CCW-1)

Dr. Crook came quickly to the end of his desired slate of tests. He announced to Willard and Davis that there were even more tests he could run, but they could not be conducted with models. They required a full-scale aircraft, "not with any view to commercial use of it, a small experimental plane in order to obtain data which he was not able to obtain in a wind tunnel." He suggested building an airplane along conventional lines but to include the channels in the wings. He could provide the specifications. Without hesitation, Davis contacted Briggs Manufacturing, and the funds to build the full-scale aircraft were released.

By April 16, 1942, Willard felt confident enough to prepare a follow-on patent application based no longer on the tabletop model, but now on the full-sized aircraft designed by Dr. Crook. Soon thereafter in August he submitted a third application, this one with Dr. Crook listed as the co-inventor. Of the twenty-six Channel Wing patents that would be issued to Willard over his lifetime, only one included a co-applicant, and that co-applicant was Dr. Louis Crook.[14] This particular application "consists of a boundary layer remover for airplanes, and has especial adaptation for use on channel or scoop type wings as illustrated in the Custer application." Of course, these two applications depended on the eventual success of the original, which had not yet been granted.

At the same time, with access to the investment capital, Willard and his team were building the first full-size Custer Channel Wing (CCW-1) in Willard's backyard workshop. Willard's oldest son Curley could not help; he had joined the Army Air Forces to learn to fly. Reed, the youngest, still attended high school. While the rest of the country was enlisting or enthusiastically building weapons of war, Willard and his team of two or three local carpenters felt good about building a unique kind of aircraft. Maybe it, too, would one day contribute to the defeat of the Axis Powers.

First, Willard focused on building the channels. For each of them, he arranged cross-sections of a conventional wing (or, airfoil) onto a semicircular frame. The diameter of the semicircle measured six feet, as Crook specified.

Next, he covered the spruce skeleton with strips of mahogany plywood that were pieced, glued, and painstakingly sanded to baby-skin smoothness. Willard also carved and shaped the laminated wood propellers by hand.

With the investment from Detroit, Davis ordered two, four-cylinder 75-horsepower engines from Lycoming, the premier aircraft engine manufacturer. He felt lucky to get them. Lycoming was laboring around the clock to supply aircraft engines to the growing U.S. war machine.

Figure 2.2 — CCW-1 Fuselage and channel under construction

Construction of the fuselage came next (Figure 2.2).[15] Willard framed it and covered it with mahogany plywood as he had done the channels. It measured only three feet wide. Somewhat surprising, the overall shape looked more like an ocean sunfish than a torpedo (the Zahm shape).

The design and construction of the aircraft required careful planning. First, when complete, the plane would measure nearly twenty feet from nose tip to tail, and the total wingspan with wing tips attached would be nearly thirty-three feet. The fully assembled plane would therefore not fit in the garage underneath his workshop, originally built for garden tractors, so it had to be disassembled for storage. In fact, the channels, fuselage, and engine mounts all had to assemble and disassemble easily. This would provide for easier transportation by flatbed truck and allow swapping out of different tail configurations for testing.

Second, unlike the configuration of Willard's tabletop model, the engines were braced inside the channels and the propellers were flipped and attached behind the engines, "pusher" style. The engines were small but heavy enough that they, with the channels, required metal braces to support them. Crook also wanted the propeller to be moveable so it could be placed halfway into the channel, or alternatively, at the back edge. So, this required devising a way to shift the engines, as well.[16]

Dr. Crook's requirements could all be met, but they would require time. As the months wore on, constant communication from Bonbright's assistants intensified the pressure on Willard. The U.S. military continued to starve for aircraft and combat materiel, they reported. They all might have an opportunity to make some money if Willard could finish his plane.

And Mr. Briggs knew how to make money. Between 1941 and the end of May 1945, the Briggs Manufacturing Company advertised, it landed 626 billion dollars in war contracts, primarily building "large aircraft assemblies, heavy bomber turrets and heavy and medium tank hulls," expanding company employment by 50%, and adding "almost a million square feet of floor space to its manufacturing operations." After retooling expenses, employee payroll, and taxes, the company cleared in profit more than five billion dollars each year in 1943 and 1944.[17]

To satisfy Bonbright, Willard hurried the final assembly of his aircraft in his tractor garage. He attached landing gear with metal braces in rudimentary fashion (Figure 2.3). He installed minimal instrumentation as this was only an experimental plane. Then he finished with paint: a primer coat of red over everything, followed by navy blue on the outside of the channels (the inside remained red), and black over the rest.

Figure 2.3 — The Bumblebee as seen from the rear

Maiden Flight

Testing the aircraft had just commenced when, on Thursday morning, November 12, 1942, Willard received an unexpected phone call.

"Daddy, it's that man from Detroit," his daughter Helen said, handing him the heavy black telephone receiver.

One of Bonbright's assistants explained that they were in Washington for another purpose, but they wanted to make a short diversion to see Willard's finished product and take some pictures.

"Okay, sure," he said.

They would be arriving within two hours.

"Okay."

And, they wanted to see it fly.

Not okay.

When they arrived, Willard recognized the three men who had sat in on his meeting in Detroit. He did not see Bonbright. They shook hands.

"We have bad news," one of them began. "Mr. Bonbright died this past June."

Willard gasped. Davis had been the middleman between him and Bonbright, and he hadn't seen Davis since April.

"I know it's a shock," the visitor responded. "He was only fifty-four. We wanted to tell you in person."

Willard silently recalled his meeting with the man and considered the present implications.

"But don't worry," said the former assistant. "Nothing has changed between the Briggs Company and Willard Custer. We're obviously still very interested and invested, or we wouldn't be here. Shall we see the plane?" he said with a smile.

Despite the masterful craftsmanship and woodwork, the fully assembled plane looked less than perfect to Willard's guests. The channels bulged like tobacco barrels strapped on by their hoops. The fuselage looked bulky. The tail's wings ("stabilizers") consisted of a flat board tacked to the tapered top edge, likely ready to be torn off even in a mild gust. And the landing gear appeared to have come from discarded wheelbarrows (Figure 2.4).

"It doesn't have to look like a bird to fly like a bird," Willard said, countering the disappointed faces. Then he remembered the ad that Frank had shown him. "A bumblebee doesn't look like a bird. It's fat and round with tiny wings that look too small to lift it. Nothing aerodynamic. But it flies." Thereafter, the first Channel Wing (CCW-1) was affectionately nicknamed, the "Bumblebee," and the awkward insect became a symbol of the Channel Wing's confident contradiction to conventional aerodynamics.[18]

Figure 2.4 — CCW-1 (side)

"We haven't flown her yet," Willard reported, trying to lower their expectations of a demonstration flight as they inspected the plane. There were strict rules about flying experimental aircraft, he explained. The federal government in the form of the Civil Aeronautics Authority (CAA, the predecessor to the FAA) required, for example, that experimental aircraft be registered and uniquely numbered, and that only CAA-certified pilots could fly them. Willard had not even registered the Bumblebee yet, let alone seen it certified airworthy. Davis had contracted with E. Kenneth Jaquieth, a certified pilot, to fly the aircraft, but Jaquieth was away from Hagerstown that week.

The men frowned.

Then Willard remembered that Jaquieth had previously been practicing taxiing with it in the adjacent field. Maybe Willard could taxi it back and forth a few times and they could take pictures in the sunlit meadow, he offered, regretting it as he said it. His visitors eagerly accepted.

So, nervous, Willard swung open the tractor gate to the adjoining field. A novice in the cockpit, he maneuvered the plane's wingspan through the gate from outside the plane and pointed it toward the meadow where Jaquieth had been practicing. Then he climbed in and started the engines. He released the brake, and the aircraft began lumbering forward over the grass.

Through the cockpit windshield he saw the short incline to the level meadow. As he would have done in the seat of his tractor when attacking an incline, Willard opened the throttle to rev the engines and gain some momentum for the climb.

The craft immediately began to feel wrong under his seat. In fact, he could feel nothing at all. Incredulous, Willard realized he was in the air! Visions of crashing, transforming his handiwork into unrecognizable splinters, flashed before him. To do so in front of these men would be perfect ruin.

Willard impulsively shut off the gas flow, stopping the engines. The plane instantly regained the ground with a sickening thud. Then, silence. His heart racing, and fearing the worst, he slipped limply out of the cockpit to inspect the damage. Surprisingly, he found nothing worse than bent landing gear. Relieved, he happily realized he would not be taxiing or anything else in this aircraft today.

As he sheepishly approached the Briggs men to deliver his apologies, they ran to greet him with enthusiastic handshakes and backslaps of congratulation. "It flies!" they exulted. Willard had piloted the first Channel Wing on its maiden flight.[19]

The experience taught Willard two things he would not soon forget.

First, the CAA regulations regarding pilot credentials and aircraft airworthiness were more than paperwork. While this short hop of a flight had not put him in much danger, he'd had a taste of how inherently dangerous experimental aircraft could be, not only to pilots but also to bystanders and property. He would not carelessly risk anyone's life to achieve this dream, so from now on, he would follow the CAA rule book for safety.

Secondly, there would be no more hurry. If Mr. Briggs' minions or anyone else outside Willard's inner circle urged him to hurry Channel Wing development, Willard now had the resolve to resist. He would not hurry for Briggs Manufacturing Company, or the war, or any other calendar or agenda. The Channel Wing would develop at its own pace.

Testing in Beltsville, Maryland

Most of another year passed before the Bumblebee was registered NX30090, certified by the Civil Aeronautics Administration in September 1943, and authorized to operate between then and March 1944.[20] During that time, Frank Kelley joined the war effort as a civilian non-combatant, went to an Army Air Force Refresher School to take an accelerated pilot training course, and got his Commercial certificate and an Instructor's rating. He then worked for the Army as a photographer in England.

In the latter part of 1943, Willard, pilot Jaquieth, and a film photographer approached a little airfield in Beltsville, Maryland. Deep forests surrounded it, secluding it and keeping its existence a secret from casual observers. The airfield consisted entirely of a beacon—allowing a pilot to spot the field from the air—and two runways, smooth with recently applied asphalt, each almost 4,000 feet long, intersecting at right angles, and laid north/south and east/west. Otherwise, a visiting pilot found no fuel tanks, no tower, and no radio antenna.[21]

The forest surrounding the field belonged to the Beltsville Agricultural Research Center. Located just north of Washington, D.C., the old air¬¬field had been appropriated and refurbished in the preceding two years by the Roosevelt administration as a potential evacuation point for the President should Washington be attacked. In the meantime, the D.C. National Guard and the U.S. Naval Reserve were using it for flight training and other unadvertised operations. Military personnel were constantly coming and going. Dr. Crook had become aware of the secluded airfield through his close work with the Navy's Aerodynamical Laboratory, and he had made arrangements to use it for Channel Wing test flights.

Willard carried the Bumblebee in on a flatbed truck. Having no fuel source at the field, he hauled that in, too, but the aircraft only held 10 gallons. Dr. Crook met the small party there and directed all of the tests. He had designed the plane conservatively, to test short take offs and landings; he had not designed it for vertical flight. And he wanted to see what difference it might make if he positioned the propeller at different points inside the channel.

In total, over many laborious visits, the airplane flew more than 100 hours (300 hours, by another account) at Beltsville. Even though it weighed 1,785 pounds with the pilot, the Bumblebee was eager to fly. Sitting still, with its flattened tires chucked, merely starting its engines would make the Bumblebee light enough to restore the tires' shape. For the first time, a fixed wing aircraft could lighten its own weight by generating static lift.

To comply with wartime restrictions, Dr. Crook never let the plane go over 60 mph or fly above the trees. It took off in 100, then 60, then 50 feet at 30 mph, sailed a few hundred feet, and then landed within another 100 feet, all before coming to the end of the runway. He measured its speed by driving a car alongside. The plane flew with and without spoilers on its con-

Figure 2.5 — CCW-1 aloft at Beltsville, Maryland

ventional wing tips. It flew alternately with the wind behind it, in its front, and coming from the side. For the first time, a powered-lift aircraft with semi-circular wings flew with better takeoff and landing performance than a conventional aircraft and with good low speed controllability (Figure 2.5).

Crook documented all the results and made design recommendations for the next generation aircraft. In sum, he wrote, "it was fully proved that the channel wing ship would easily clear a 50-foot obstacle within a 50-foot take-off run, and could entirely disregard wind direction in either landing or take-off." The take-off run measure was significant. By contrast, a conventional twin-engine aircraft, such as the modern Cessna 310 or Beech Baron, being twice as heavy as the CCW-1 but having two to three times the wing area, still require a much higher liftoff speed and over 1500 feet of runway ground roll to clear a 50-foot obstacle on takeoff. This is primarily because the powered-lift Bumblebee could generate 2.4 to roughly ten times the lift capability as the conventional wing.

Willard had previously learned how critical motion pictures could be to document his plane's performance, so in Frank Kelley's absence another photographer recorded the flights. He had to stand close to the runway because of the encroaching woods, and so had difficulty keeping the airplane in focus as it made its long hops. Even though his films were grainy, they captured the airplane's capabilities sufficiently to provide demonstration value for anyone who could not attend in person.

Reuben Davis kept the treasurer's office at Briggs updated. As directed, he also spoke with some representatives of the office of Army Air Training who frequented the Beltsville field, and requested that they, in turn, contact the Army's Air Materiel Command. The new Briggs treasurer wanted the Army Air Force to see this plane. (The U.S. Air Force did not become a separate service branch until June 1, 1947.) Briggs Manufacturing hungered for another military contract.

Responding to the invitation on December 17, 1943, Brigadier General William E. Gilmore came to Beltsville from Wright Field in Dayton, Ohio to see the Bumblebee in action. A blunt, no-nonsense man, Gen. Gilmore had responsibility for all aircraft procurement for the Army Air Corps. The Bumblebee so impressed him that he ordered it put on the list for Wright Field analysis, testing, and evaluation as a candidate for Army Air Corps development. Gen. Gilmore even invited 72-year-old Orville Wright—a long-time, personal colleague who happened to be in Washington on the same day—to come see the unique plane.[22]

Willard, of course, celebrated the news, but cautiously. On December 5, and quite by surprise, he had received a second patent rejection notice. This time, the examiner had discovered that nearly six years prior, someone else had already patented elements that appeared to be similar to Willard's airplane design. Because Willard's invention failed to "patentably distinguish over the art" (which is Patent Office parlance for "someone beat you to it"), it could not be approved. If Willard couldn't resolve this issue, it wouldn't matter what test results came out of Wright Field. The Bumblebee could well be the first—and last—Channel Wing aircraft ever to fly.

III The Experts 1944 - 1948

With further improvements in engine design, the aircraft, if properly made, could be designed to rise vertically.

- Don Young, US Army Air Force Engineer

Besides, I am not convinced that plaintiff's will actually climb vertically. I don't see how it could.

- Edwin L. Reynolds, US Patent Office attorney[1]

Early in 1944, Garvey and Willard sat in leather chairs at the end of a long, mahogany table in Garvey's law office, discussing the latest rejection letter from the Patent Office. Willard suspected the examiners, unwilling to admit that he could know more about the Channel Wing's aerodynamics than they did, had latched onto the first flimsy excuse they could find to deny him his patent. Garvey tried to explain the issue in terms that Willard, uninitiated in patent law jargon, could understand.

"The Patent Office will not grant a patent to two men who separately invent the same invention," he began.

Willard knew that.

"The patent examiners have come across a patent by a man named Matthias Henter," Garvey said, passing Willard a copy of the patent to which he was referring. It showed a diagram of an aircraft that was little more than a single wing. The fuselage was a bulge in the middle of the wing, so that it resembled today's B-2 Stealth Bomber.

"Notice the propellers," Garvey said as he pointed to the diagram.

"They're pushers," Willard observed, studying three propellers mounted on the back of the wing. "That's not a Channel Wing! That's just a wing with pusher propellers! How can they confuse that with mine," he said, as a statement more than a question.

"That's right," Garvey agreed. "That's why, when I did a prior patent search before submitting your application, I didn't call you, and say, 'Sorry, Willard, someone else has already invented your airplane.' But the examiners don't see it like that. They see an aircraft with propellers mounted at the back of the wing to increase lift. What we must show them is that yours is different—or, if not different, then better."

"Mine is different and better," Willard asserted. "And after four years of arguing, they know it, but they still refuse to admit it."[2]

In May of 1944, Garvey proposed to resolve the Patent Office examiners' objections by further changing the wording in Willard's patent. Garvey, Willard, and Dr. Crook visited the Patent Office to discuss the matter in person. To document the points made in that discussion, Garvey recorded, in part, the following notes:

So far as the applicant has been able to ascertain, he is the pioneer in the construction of a wing capable of exerting static lift. That he has solved this problem is evidenced by the fact that a full sized plane has been completed and subjected to numerous tests. Motion pictures of this plane in flight were shown in Richmond. In addition, literally hundreds of wind tunnel tests have been made, principally by the noted authority, Louis H. Crook. An Affidavit drafted along the lines requested by the Examiner is attached hereto. In this Affidavit Professor Crook unequivocally asserts that with the Custer plane of the present invention, a new and unexpected result is obtained.[3]

Willard and Garvey felt better after the meeting and hoped for a positive outcome. Surely Dr. Crook's position on the matter would hush any dissent by these men of lesser educational degree. Consequently, on June 17, 1944, Willard filed a fourth patent application. Entitled, "Multiple Propeller Wing Channel," this one "is a continuation in part of my applications...filed April 16, 1942; and ...August 28, 1942, and its object is to provide channeled wings substantially as shown in the above applications in conjunction with power means at both ends of the channel..." The innovation was to put a propeller at each end of an elongated channel.

The application was filed eleven days after the Allies invaded France,

D-Day, June 6, 1944. The war had taken a turn for the better, and Willard felt his patent prospects had, as well.

Army Air Force Test Report 5142

Gen. Gilmore had seen the Bumblebee fly and had promptly ordered it to be put on the list for Army testing at Wright Field in Dayton, Ohio. Accordingly, in June 1944, Willard left Lula at home for a few days with their youngest child, Reed, now nearly seventeen years old, and his sisters, both in business school and husband-hunting. Willard traveled to Wright Field to provide a new model for the Army Air Force tests of his channel-and-propeller concept. The model consisted of a 2-foot-diameter channel with a matching propeller. He also supplied copies of a few of Dr. Crook's reports. Willard was ushered into the facility, shown around, and briefed on the testing protocol. He met the engineers who would be conducting the tests to measure the channel's lift at various airspeeds in a 5-foot wind tunnel. He sensed the chilled skepticism of the technicians. Worse, the lead aeronautical engineer for his project was one of the most skeptical of them all, Donald W. Young, an Aerodynamic Specialist.[4]

At that time, operations at the Materiel Command at Wright Field were still running at full capacity to support the war effort. Working around the clock, the aeronautical engineers had been testing, evaluating, and awarding production contracts for all of the U.S. Army aircraft deployed since the beginning of the war. Requisitions for attack and transport aircraft continued, coming in faster than the Materiel Center could fulfill them. The testing and acquisition process had been abbreviated and streamlined as much as possible, and the drive to deliver for the war effort continued without reprieve.

In this environment, Don Young conducted his tests of the Channel Wing from June 6 through July 3, 1944. He then recorded his test procedure and results in a thirty-four-page report (AAF Report 5142) dated September 5.[5] The report included two sketches of the channel/propeller arrangement, eighteen graphs of plotted test data, and eleven photographs. In the back of his report, a list of four references cites a report by Dr. Crook, entitled, "Lift Forces on Custer Model Scoop with G-398 Airfoil Section."

Young's report presents his test results, conclusions, and a recommendation. Among the test results, the report includes the following. First, "the take-off speed of the Custer airplane with channels and pusher propellers is 36 mph. This is in comparison to a take-off speed of 51 mph if the channel section were replaced by a normal wing section..."

Second, "In order to evaluate the device under test as a means of producing static lift, the wind-tunnel results were compared with data of a representative helicopter. This comparison shows that the static lift of the device under test is 8.4 pounds per horsepower, while that of the helicopter is 15.2 pounds per horsepower."

The conclusions of the report are as follows:

1. Upon the basis of the results of the present tests it is concluded that the increment of lift due to propeller operation on the device under test is caused by the slipstream velocity [Willard's "speed of air"] and is of the same nature as that existing with conventional wing-propeller arrangements.
2. The magnitudes of the lift increments obtained show that the present device is markedly inferior to the helicopter in producing static lift but is superior to conventional wing-propeller arrangements in producing both static lift and lift when forward velocity exists.
3. The present device does not show sufficient promise of military value to justify further development by the Army Air Forces.

The second conclusion, that Willard's design was "superior to conventional wing-propeller arrangements in producing both static lift and lift when forward velocity exists," came as no surprise to anyone in Hagerstown or Cabin John, as Dr. Crook's experiments had produced similar results. What was disappointing, however, in light of the positive findings, was the third conclusion that it showed insufficient value to the military.

Finally, the report's recommendation left the matter unsettled. "It is recommended that no further consideration of the present device be given until improvements, expected by the inventor, have been verified."

At least the Army had left the door slightly ajar.

Of course, the report disappointed Willard. He suspected the test results had been intentionally underplayed, prejudiced toward Sikorsky's helicopter, which had recently been tested and was already in production.

These two men and their flying machines could not have been more different. Igor Sikorsky had been born in Kiev, Russia, ten years before Willard's birth. His father was a doctor and a professor of psychology; Willard's father was a blacksmith. Igor had attended the Naval College in St. Petersburg and the Mechanical Engineering College of the Polytechnic Institute in Kiev; Willard didn't finish high school. Igor had built his first helicopter in 1909 when Willard was ten. He became a pilot, flying his own aircraft of conventional design, and migrated to the United States in 1919. Four years later he started his first company, making conventional aircraft, and sold the company four years after that. He got his first helicopter patent in 1931 and flew the first successful one in 1939. Military contracts followed, and in 1943 his helicopter went into production.[6] Willard's experience was very different.

Sikorsky's helicopter, as different as it may look from a conventional airplane, obtains lift in the same way. The rotor of the helicopter is shaped like an airplane wing (airfoil). And, like an airplane, it generates lift by moving the airfoil through the air. The airplane wing moves in a straight line down a runway, while the helicopter rotor moves in a tight circle. In both aircraft, lift is obtained by moving the aircraft. In contrast to both of those, a Channel Wing aircraft moved the air.

Meanwhile, back in Maryland, Crook completed his tests in November. Willard disassembled the Bumblebee and stowed it in his workshop garage. He and Crook made plans for the next generation Channel Wing airplane, working toward a full-size model that could take off and land vertically.

Then Garvey got news from the Patent Office that the examiner's rejection was final. Willard's patent application would not be accepted, even in its revised form. Willard's former feeling about these "smart boys" returned. They were, he felt, simply unwilling to be proven wrong, unwilling to learn something new from him.

The U.S. Patent and Trademark Office has an appeal process should an inventor insist that an examiner's rejection is not justified. Garvey immediately set in motion an appeal to the Patent Office's Board of Appeals. In February of 1945, he and Willard appeared before the Board and offered to further rewrite the claims in the application and argued again that Willard's design was different from any prior patents. Garvey showed the Board the film of Willard's model demonstration. And then he drew on the recent Army Air Force report:

> The airplane which has been built and flown, modifies the wing structure herein shown but basically it remains the same in that the troughs or channels are mounted on the opposite sides of the fuselage and the propellers are arranged in such close proximity to the aft edges of the channels that there is no appreciable escape of air between the propeller discs and the trailing edges of the channels. The model of

> the wings constructed in accordance with the teaching of this invention has been tested by the Army Air Forces at Wright Field, Dayton, Ohio and in Report No. 5142 dated November [actually, September] 5, 1944, the following conclusion is drawn.
>
> 'is superior to conventional wing propeller arrangements in producing both static lift and lift when forward velocity exists.'

Finally, Garvey concluded his appeal with the following assertions:

> 1. The invention here at issue is of a revolutionary character.
> 2. Tests made by applicant on model, as well as on a full size airplane and on wings per se, are indicative of a novel phenomena [sic] in aeronautics.[7]

The Board would study the issue some more, it said, but it offered no promises on a decision date.

On May 29, 1945, Don Young, the Army Air Force's lead tester of Willard's model at Wright Field, visited Willard at his shop in Hagerstown. The two had kept in touch, Willard insisting he had made some improvements to his design, and Young promising to review them when they were ready. The war in Europe had eased (Germany surrendered in May), and on May 25 Young received the orders to go to Hagerstown. His purpose: to "determine the resultant force in pounds/HP obtainable with Custer Channel Wing" based on Willard's changes. Willard determined he would score well in comparison to the helicopter's performance.

"The model," Young wrote in his report, "consisted of a U-shaped channel made of sheet metal and having a radius of 11.5 inches with straight sides extending above the center line 6.5 inches. The length of the channel was 11.5 inches. The propeller, located 0.5 inch ahead of the trailing edge of the channel, was supported on a shaft having three bearing supports."

Most novel in the configuration, however, was the conventional wing placed directly behind the channel wing, in the air stream produced by the propeller. This auxiliary wing measured 36 inches wide and 24 inches long. The two wings were mounted to the same base. Young conducted the tests.

Based on the data obtained in the tests, Young documented the following conclusions:

First, "A 55 percent increase in static lift of the Custer wing was obtained by placing a straight auxiliary wing aft of the channel wing."

Then, "This arrangement makes a favorable comparison with the helicopter based upon the power data calculated for the propeller."

Finally, "The arrangement indicates sufficient merit to warrant further investigation."

He ended his report by recommending that the Army Air Forces test "the arrangement" more completely. He even sent Gen. Gilmore, the head of the Materiel Division and the first to recommend tests of the Channel Wing, a copy of this memorandum.[8]

Willard's efforts with Wright Field might yet pay off. Don Young's appraisal of the Channel Wing had finally taken a decidedly positive turn.

But the Patent Office's view had taken no such turn.[9] On June 15, 1945, the Patent Office's Board of Appeals dropped its bombshell decision: the patent was denied. In August, the United States brought out its secret weapon and dropped it on Hiroshima and Nagasaki, and Japan capitulated soon afterward. The world war had ended, but Willard's patent fight had not. Willard would not capitulate.

He knew that the Briggs Manufacturing Company, like any investor, would withhold its funding if he lost his patent fight. His income over the previous eighteen years had always been tentative, and he had accepted that insecurity as a temporary liability in an inventor's existence. It caused him to provide lunch boxes to his children and point them downtown to find jobs when they graduated from high school, rather than bringing them into "Willard and Sons, Inc." They needed a sense of financial security—a comfort he could not provide them, a principle he could not teach them; at least, not yet.

But the money mattered only so much.

The supremacy of observable facts, the documented results of the Channel Wing's performance—that is what mattered most. Willard believed that, when honest patent engineers considered the objective facts, they would do their jobs without prejudice and issue him the deserved patents. How could it be any other way?

With no alternative, then, on December 10, Willard and Garvey filed suit in the District Court of the United States for the District of Columbia; Willard R. Custer, Plaintiff, versus Casper W. Ooms, Commissioner of

Patents, Defendant; Civil Action Number 32002, Complaint for Issuance of United States Letters Patent.

It had come to this.

But the docket bulged, and Willard would have to wait again.

Army Air Force Test Report 5568

In 1946, Frank Kelley returned from Europe, where he had been a photographer for the U.S. Army. He met soon after with Willard.

"How's the family?" Frank wanted to know. He and his wife Elizabeth had no children.

Lula was well, Willard reported. Curley served stateside with the Air Force, and Reed had enlisted with the Navy, now serving at Camp Peary, near Williamsburg, Virginia. Both boys were safe. The girls, Helen and Vivian, were finishing at school—one at Hagerstown Business College, the other at the University of Maryland—and occasionally organizing his files, taking correspondence, filling out his forms.

In February 1946, Don Young commenced his final round of tests at Wright Field, officially designated as "Test of Two Custer Channel Wings Having a Diameter of 37.2 Inches and Lengths of 43 and 17.5 Inches (Five-Foot Wind Tunnel Test Number 545)," or simply, AAF Report 5568. It involved a much more extensive test plan and "53 different model configurations...on two different length channels, with two and three blade propellers of various planforms [i.e., contours] and blade angles. Tests were also made to determine the effect of placing auxiliary wings of various sizes and arrangements aft of the propeller-wing combination."

Conducting the regimen required more than three months, from February 6 to May 22, 1946. The sixty-nine pages of the report, including eleven tables represented in twenty-five graphs and twenty photographs (Figure 3.1), asked and answered every remaining question Young could imagine regarding the Channel Wing.[10]

At the end of the report, Young concluded the following:

1. The shorter channel is superior to the longer one.
2. The two-blade propeller is superior to the three-blade.
3. While auxiliary wings behind the propeller increased lift, they are a liability once forward movement commenced. The advice is to not include them in a production aircraft.

Figure 3.1 — Channel Wing testing at Wright Field, Dayton, Ohio

Yet, despite all of that, the report's official recommendation is simply put: "None."

The result of all that effort, the tests and retests, the laborious documentation of positive results, all to result in nothingness. Futility. "None!"

When he thought things could not get any worse, Willard got a phone call at the house.

"Daddy, it's for you," his daughter Helen said, handing him the receiver. "It's Mr. Garvey."

"Yes," answered Willard. "What is it now?"

"Are you sitting down?" Garvey asked, and waited for Willard to say yes. Garvey had something important, and like the accomplished trial lawyer he was, he couldn't resist being theatrical.

"I was doing some research at the Patent Office the other day, to see if any recently issued patents bore any similarity to yours, and what do you think I found?"

Willard bit. Now Garvey reeled him in.

"I'm holding in my hands a copy of a patent for an 'Aircraft' which, it says, 'can be more nearly dependent upon the propellers rather than the forward travel of the airplane along the ground to effect the lifting of the airplane from the ground, and the preferred construction is such that the airplane, when the propeller speed is sufficient, will rise or descend in a generally vertical direction—'"

"What patent is this?!" Willard interjected.

"Wait, listen to this," Garvey continued reading. "'When the propellers are turning from about 2,500 to 3,500 R.P.M. a low pressure area of tornado proportion is created over the wing sections which can be increased or lessened by varying the R.P.M. of the propeller.' And, 'At about 1,500 R.P.M. the wing sections cease to weigh anything as the atmospheric pressure under the wing sections exceeds the weight of said sections.'"[11]

Garvey stopped reading. Willard had stopped breathing.

"Willard?"

"I wrote that," Willard said, hoarsely.

"I know you did," said Garvey. "And don't you want to know whose patent this is?" Without waiting for Willard's answer, he announced, "It's Howard Bonbright's!"

Willard's lungs collapsed.

"Assigned to Briggs Manufacturing Company!" Garvey continued. "Willard, while you were keeping him in the loop about your troubles with your patent, Mr. Bonbright was lifting your very words and putting them into his own patent application—it appears to me—and while we were wrangling with the Patent Office over their perceived similarities between your application and Henter's, the Patent Office approved Bonbright's patent—which uses your word-for-word verbiage—just a few months ago!"

A tornado of questions twisted inside Willard's brain. Bonbright had betrayed him before he died? How had Bonbright gotten a patent issued using some of Willard's own text, while the Patent Office refused to issue Willard's patent because of some obtuse similarity with Henter's? How could Bonbright have done this to him? Was the Patent Office in collusion with him? Could he trust no one to be honest?

Garvey listened to the stunned silence as long as his flair for drama demanded. Then his friendship for Willard and his fear for Willard's heart overcame him, and he said with a smiling voice, "But don't worry, Willard. I've taken care of it."

"What do you mean?" was all Willard could croak as his memories of

Bonbright swirled.

"I've called the Briggs attorneys and the Patent Office, and—in my uniquely persuasive manner—gotten them to agree to the following: You, Willard, will forget you ever heard of this patent if they, Briggs, will bury it with the late Mr. Bonbright, God rest his soul. And, they will release you forever from your agreement with them. Meaning, they will say good-bye to all the money they've given you, and you will never have any obligation to share any of your profit with them, no matter how rich and famous you become."[12]

With this agreement, irrespective of the outcome of the pending patent trial, Briggs Manufacturing Company's sponsorship of the Channel Wing effort ended. Willard's immediate response was to install iron bars on every window of his workshop, and install locks on every door, file cabinet, and pair of lips in his family and board of directors. And he vowed he'd never let anyone double cross him again.

Willard Custer v. Caspar Ooms

Due to a full federal docket, the trial on the denial of Willard's patent application, entitled "Aircraft Having High-Lift Wing Channels," did not start until Thursday, May 15, 1947, eighteen months after Willard had filed suit.

The presiding judge, Associate Justice Edward Curran, had just been appointed to the federal bench by President Truman the preceding October, but he was not seated until February. In 1949, two years later, he would preside over the trial and sentencing of Mildred Gillars ("Axis Sally"), who would be found guilty of broadcasting Nazi propaganda during the Second World War and sentenced to serve ten to thirty years in prison. But in May 1947, Curran had occupied this bench only three months when Willard came before him.

Representing the U.S. Patent and Trademark Office, Edwin L. Reynolds, Esq. had started with the Patent Office as an assistant examiner right out of college twenty-five years prior. In 1927, he had gained a law degree from George Washington University, and he had been in his current position of law examiner since 1940. At forty-six years old, he was a rising star. Three years later, in 1950, he would become a colleague of Garvey's when the two of them would teach at Georgetown University Law School as adjunct professors of patent law. In 1961, President John F. Kennedy nominated him to the position of First Assistant Commissioner of Patents where he would serve until he retired in 1969.

Joining Willard and Garvey at the plaintiff's table were Garvey's business partner, Joseph A. O'Connell, Esq., and Frank Kelley. Frank, no longer the president of Willard's company but still its official photographer, manned the film projector.

Because Garvey represented the plaintiff, he introduced the judge to the facts of the case.[13] He began by describing the working model that Willard had produced, and how the patent application in question was based on this model. He recounted how the application had been originally rejected on grounds it was "inoperative," and how this had been reversed after the examiners viewed the film of the model in flight. And he described how the examiners had rejected the application again on the grounds that it failed "to patentably distinguish over the art."

Then Garvey told the story of the investors from Detroit, Dr. Crook's tests, and the construction and flights of the full-scale aircraft at the Beltsville airfield. Finally, Garvey prepared to explain to the judge the reasons for the Patent Office's rejection and the issues on which the judge would be required to rule.

"I would like to have the claims explained to me," Judge Curran said. "If you can describe them to me in the English language, not that of an inventor or engineer. It is a little difficult for me to understand it."

In plain English, Garvey explained the reason Willard's patent application had been finally rejected after six years of debate and wrangling.

He explained that in 1931 and subsequent years, Mathias Henter had been issued multiple patents on an airplane design he called "the flying-wing type." His aircraft consisted of a single airplane wing with the fuselage integrally molded into the airfoil. His design allowed any number of propellers to be mounted behind the wing in "pusher" configuration, that is, at the back of the wing.

Two diagrams from Henter's patents are reproduced here. Figure 3.2 is a cross-section of the wing showing how the propeller is mounted. The propeller (labeled number 1 in the diagram) is positioned within the profile of the wing (labeled 4) in such a way as to preserve the wing's aerodynamic shape. The profile is preserved by means of a semi-circular depression (6) in the wing just ahead of the propeller. When the propeller turns, said Henter, it draws air over the top of the wing resulting in more lift than is afforded by conventional craft.

The second diagram (Figure 3.3) had been enlarged and mounted on a blackboard in the courtroom. A top-down view of an entire airplane in one of Henter's patents, it shows the single large wing (30) and three propellers mounted behind it. The middle propeller is not couched within a wing

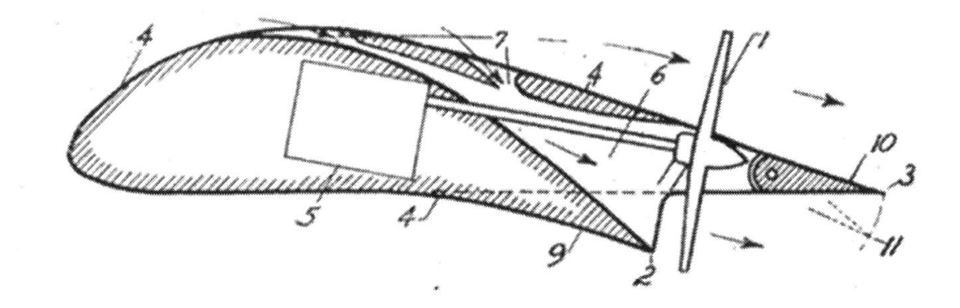

Figure 3.2 — Cross-section of Henter's wing

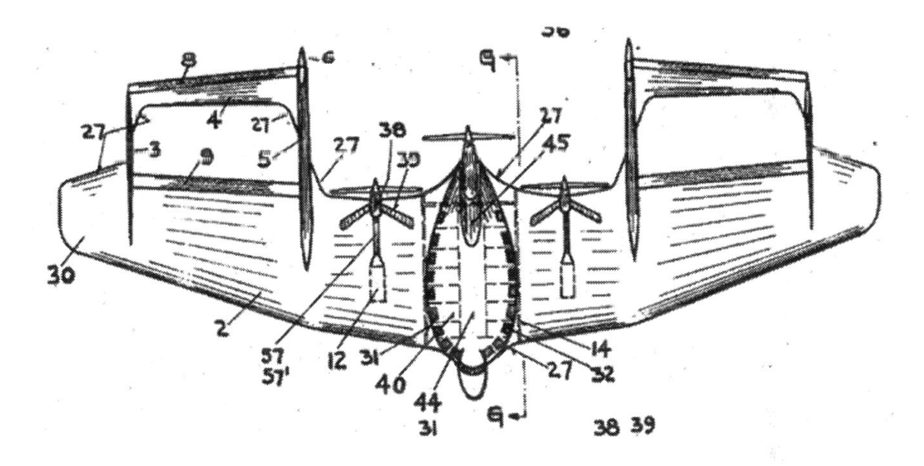

Figure 3.3 — Top-down view of Henter's wing

depression as illustrated above, but the two side propellers are. In addition, this figure shows what Henter called an "outrigger" (5) and a "stabilizing surface" or "fin" (3), both of which are uprights that support the tails (8).

Also in the court room, a drawing (Figure 3.4) from Willard's patent showed some of the components of his application that are similar to Henter's. For example, the propellers (15) are mounted behind the wings (8, though not as clear in the head-on view as it is in the top-down view), and an upright (9) on the wing tip that Willard calls a "fin." Willard's design also includes "baffles" (21) which round out the corners formed by the intersection of the wing and the fin, and the wing and the fuselage (5).

Garvey fairly presented the two aircraft designs and the Patent Office's position that the two designs were so similar that an examiner could not distinguish one from the other to allow Willard a separate patent.

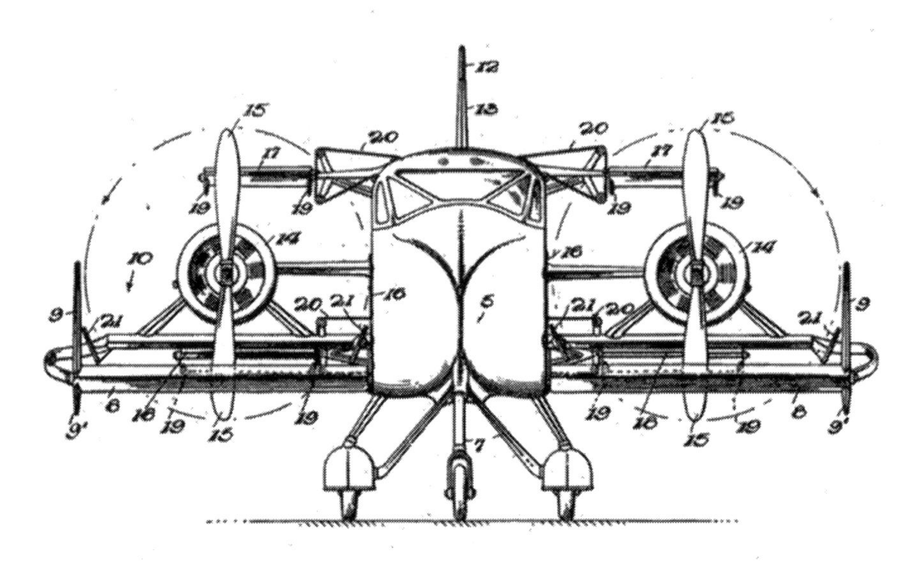

Figure 3.4 — Willard Custer's patent diagram

"All right," Judge Curran said as Garvey wrapped up. "Now my question is this: Does Custer have the same thing in what he calls 'the fin' as what Henter had with what he called an 'outrigger'?"

Mr. Reynolds, representing the U.S. Patent Office, jumped to answer.

"They are not of identical construction, no. But I think the term 'fin' applies to both. Henter's does not rise vertically, but they are both airplanes."

"I know they're both airplanes," the Judge snapped. "But they are set up differently."

"Yes, of course they are," Reynolds agreed. "But the question is, does the language of one claim distinguish from the other structure? That is the issue you have."

"But Henter does not call it a fin," the Judge pointed out. "Henter calls it an outrigger. He says that the outer tips of the wings, the upright and surfaces or fins—"

"But as you can see," Reynolds interrupted, "the structure is exactly the same! 'Fin' is a structural term, not a functional term."

The Judge nodded to Garvey. "Go ahead."

"Your Honor," Garvey continued, "we are prepared to go into this in considerable detail through the mouths of experts who know far more about it than I do, and certainly far more about it than even the inventor. After all, as Your Honor knows, it is not necessary for the inventor to know the scientific principle of his invention. It is only necessary that he had this device and that it works. If it works, that is the important thing."

"But have they both got the same thing?" Judge Curran repeated. "I know they both have airplanes," he said, darting his eyes at Reynolds, "but have they both got the same airplane?"

"Oh, no!" Reynolds interjected. "They haven't got the same airplane, of course!"

Reynolds, despite admitting that the two aircraft designs were very different, would not budge from his position that they were too similar in structural parts and wording to distinguish between them. According to the Patent Office, Henter's use of a depression ahead of the pusher propeller, and its placement between the outrigger and the fuselage, constituted as much of a channel as did Willard's use of a pusher propeller walled between the fuselage and the fin.

Garvey, on the other hand, tried to argue that the functionality of the two designs was very different. Henter mounted his propeller in relation to his wing in order simply to increase lift, but Willard's design was intended to produce static lift, a vertical ascent, Garvey said. Reynolds, however, argued that the functionality of the design was not patentable; rather, it was the structure that is patented.

"Besides," Reynolds added, "I am not convinced that plaintiff's will actually climb vertically. I don't see how it could."

Following a recess for lunch, Garvey called Willard to the stand for the preliminary oath and introductory questions. Then, Garvey had Frank show the film, the one where the model rose vertically from the table. Garvey quizzed Willard about the model's design at certain points during the film.

After the film, Garvey asked Willard about the full-scale airplane—the Bumblebee—its design, and its flights at Beltsville. He also asked a few questions about the Channel Wing's performance during tests conducted by the Army Air Force.

The stage was now set to contrast the Channel Wing's performance with that of Henter's wing.

"Mr. Custer," Garvey addressed him, "did you ever make any effort to determine whether or not a device was ever constructed in accordance with this Henter patent we are now discussing?"

"Yes, sir," Willard answered. "I went and talked with Mr. Henter."

"When was that?"

"In 1946," answered Willard. He and Garvey had rehearsed these lines.

"Do you know," Garvey asked, "whether or not that plane was ever flown? Whether a model was ever made in accordance with the structure shown in this Henter patent?"

"I think that is hearsay," Reynolds objected.

"I will sustain the objection," Judge Curran replied.

Garvey reworded his question.

"Do you know," he asked Willard, "of your own personal knowledge, whether or not the Henter plane has ever been flown?"

Willard replied, "I know he said it hadn't."

"I object to that!" Reynolds repeated.

Judge Curran replied, "Objection sustained." Then looking at Willard, he offered, "You don't know."

"No, sir," Willard disagreed, and he rephrased it himself. "I have never seen the Henter plane built nor seen one built like it."

Garvey's point had been made. He was focusing on actual airplanes and demonstrated results, not patent claims or interpretations of patent drawings. The fact that only Willard's aircraft had actually been built and flown might carry weight with the judge.

After more discussion on Willard's design as it differed from Henter's, Garvey delivered Willard up to Reynolds for cross-examination.

"Now, referring to the Henter design," Reynolds began, turning toward the easel and pointing to the number 3 on the enlarged diagram from Henter's patent, "is it your testimony that the part of it which lies above the wing could not properly be called a fin?"

"Yes, sir," replied Willard.

"Will you explain why?"

Willard replied, "It would not be close enough to the propeller to give us the action we need for lift."

"That has nothing to do with whether it is a fin or not, has it?"

"Well, according to the normal name," Willard explained, "that is not a fin; it is not termed a fin, it is not called a fin by the industry."

"Then how would you design a fin?" Reynolds inquired.

"In my opinion, it has a very thin surface in order to be as close to the propeller as possible, very close," described Willard. "Otherwise, it would not cause the necessary condition."

"You understand," corrected Reynolds, "I'm just talking about the word 'fin' now. And my question is whether that thing—" Here Reynolds pointed again to the diagram, "Whether that thing is a fin or not. I'm not referring to any propeller at the moment."

"I would not think so," repeated Willard.

"Well, then," responded Reynolds, "do you think that this part here," pointing to the number 5 on the chart, "this part which projects above the wing is a fin, or not?"

"I don't think that is a fin," Willard repeated. "I should say that could not be a fin."

"Now, then," Reynolds asked, "will you explain why it is not a fin?"

"Well, in my opinion," Willard answered, "as I said before, a fin would be something very thin to serve a certain purpose, a purpose which I needed."

"Then you think it is too thick to be a fin, is that right?"

"Well, it is too low; it is not too thick." Willard imagined the Bumblebee. "We use the side of the fuselage for the same thing, so long as the propeller runs very close to it, and gives a channel."

"I want you to forget about what it is in your design!" Reynolds repeated. "I am just asking you about the word 'fin.' The word 'fin' is used to define a structure such as this, is it not? It has a definite meaning, hasn't it?"

"Probably so," answered Willard.

"You don't know what that meaning was?" asked Reynolds.

"No," answered Willard.

"You don't know what a fin is, is that your answer?"

"I know what I use it for," Willard answered.

"You have given us the only reason you can think of why you consider that is not a fin, have you?" begged Reynolds.

"In my opinion," Willard countered, "that is an outrigger, as the gentleman says who invented it. It is there to support the tail."

Willard was right; Henter had called it an outrigger.

"Well, this plane of Henter's has a fuselage, does it not?" Reynolds asked.

"Yes, sir."

"Would you say it is provided with wind channels, or not?"

"As I know a channel," Willard replied, "it is not."

"Why is it that the space between the fuselage and outrigger is not a channel, in your opinion?"

"Because," Willard replied, "in my opinion that is just a pusher aircraft. I have got to nail that close and tight. In my opinion, that is just a pusher aircraft."

"But is it a channel? Yes or no?" rebounded Reynolds.

"No, sir," Willard stood firm, "in my opinion it is not."

"Then how do you define a channel?"

"I define a channel exactly as I show it in my application," Willard answered. Now Garvey became involved, too, as questions about channels, outriggers and fins continued unabated for more than an hour. Finally, Judge Curran ended all of the questioning, and summarized the issue succinctly:

"The situation with the Patent Office, as I see it, is that they have issued a patent to a Mr. A wherein he used the terms 'wing channels, outriggers, fins, propellers' and so forth. Now along comes Mr. B and uses the same terms. It might be that the idea of Mr. A is different from Mr. B; it might well be that Mr. B's structure is much improved and much better than Mr. A's. But the fact remains that a patent has been granted to Mr. A. Now, can Mr. B come along and obtain a patent on his invention when the Patent Office claims that the disclosures of Mr. A are practically the same as the disclosures by Mr. B. Isn't that the point?"

"That's exactly our position," hurried Reynolds.

"If the disclosures are practically the same," argued Garvey, "the inventions are not."

"That is for me to decide," Judge Curran answered. "I am not deciding it yet, you understand. I have taken no position about it."

It was 3:50 p.m. and the litigants were exhausted. Judge Curran adjourned the proceedings until 10:00 the next morning. As usual, Willard felt he had scored no points against the Patent Office. They had made up their minds years ago when they first met him. They wouldn't admit to vertical lift if it took them to the moon. He might as well talk to a brick wall.

The next day, Friday morning, May 16, found both parties reassembled before the Judge. Garvey had but one more voice to offer to the argument, and then he would rest his case. He called his expert witness to the stand.

Professor Crook had lent his scientific credibility to the support and development of the Channel Wing. The benefit of the aeronautical expertise he had brought to the testing and engineering designs of the aircraft in its infant stage could not be exaggerated. He was still under contract and doing work for Willard's company.[14] Given this and his visits to the Patent Office to explain Willard's invention to the examiners, one might have expected that Garvey would have called him as his expert witness. But Dr. Crook did not appear.

"Please state your full name, age, residence and occupation," Garvey said after swearing in his witness.

"Don W. Young; Dayton, Ohio; Aeronautical Engineer, Aerodynamic Specialist, employed by the Army, Field Command, at Wright Field."

"How long have you been so employed?" Garvey asked.

"About nine and one-half years," Young answered.

Garvey asked Young to describe the tests of Willard's device that he had conducted over the past few years. And after retelling some of the details of those tests, Garvey brought him to the results.

"Is it possible for you to state the result of all those tests made, whether or not you got greater lift by this arrangement than with the standard airplanes? Just describe that in your own way."

"Those series of tests were to tell us what that device would do. From other data we had from other normal airplanes, which was our reference data, we were able to tell that this would produce approximately three to four times the lift with no forward velocity such as the average airplane would have."

"By 'this' you mean the Custer scoop?" Garvey clarified.

"Yes, that is right."

"Now, Mr. Young, were you here in the court room yesterday when the motion picture was exhibited of the Custer model?"

"Yes, sir."

"Based on your experience and education," Garvey said, "it is your opinion that that model did fly vertically as represented in that picture yesterday?"

"Yes, sir," he replied.

Then Garvey asked, "You think that entirely based on the Custer device, it can fly vertically?"

"Yes, sir," he said, "when considering that model."

"Now," Garvey asked, "do you think it would be possible to construct a Henter device to fly and do the same that that model did as you saw yesterday?"

"I don't think it would."

"Do you think it would be possible to get vertical rise from this Henter device as you saw demonstrated in the Custer model yesterday?"

"No, sir," Young answered.

"Why wouldn't it be possible?" asked Garvey.

"The primary way you get vertical rise in the Custer device is from the channel wings, and the lift that you would get from this Henter device," Young said, nodding at the patent figure still posted in the court room, "would not be sufficient to give you the lift so as to overcome the weight and give you a vertical rise."

Garvey continued, "Based on the tests which you have made of the Custer models which were brought to you at Wright Field, is it your opinion that under proper conditions, that a plane may be constructed which will rise vertically?"

Young replied, "It is my opinion that a device of this type, such as the Custer design, with present aircraft engine weight and results of our tests, for a design to be made whereby the lift of the aircraft would be approximately 75 percent of the weight of the aircraft. With further improvements in engine design, the aircraft, if properly made, could be designed to rise vertically."

Willard's heart soared when he heard those words. He knew Dr. Crook had caught his vision, but Dr. Crook had never said it this way in this context to this audience. Willard had been working so hard to get someone in the government to understand. The examiners seemed obstinate in their refusal to open their minds, and the Army Air Force had been slow to learn and then impotent to act. Now, on hearing this from the Army's aeronautical expert, Willard had to hold back tears.

Young was doing well. Smart and articulate, he knew his field of study, from the academic and theoretical aspects to the practical and testing concerns. Best of all, he knew the Channel Wing. Judge Curran seemed to like him, too.

But then Mr. Reynolds counter punched. He reminded the Judge that the recommendations published in Young's studies were anything but positive.

"Let me explain this," Young began. "I happen to be the author of both of these reports. When you are working for someone, you have to write with the viewpoint that will carry the authority of those over you, and necessarily the author may not be in entire agreement with the contents of the report. In other words, he has to write from the viewpoint of the higher authority."

"You mean to say you wrote a report to express your opinion when it was not your opinion?" Judge Curran summarized.

"Sometimes you have to," answered Young. "There is a way you can get around it by saying, 'The opinions represented here are not necessarily those of the author,' but I can assure you, when you do that, you are adding fuel to the internal fire."

Young then recited from memory: "'It is recommended no further consideration of the present device be given until improvements expected by the inventor have been made.' Later we made some more tests which dealt with all these other factors. But in 1944, it was recommended that 'no further consideration of the present device be given until the result expected by the inventor has been verified.'"

"But, why?" asked the Judge.

"Because," explained Young, "it did not appear to have sufficient merit at that time. We didn't think it would justify any continued work on it."

"Why?" asked the Judge again.

"We were interested in a short take-off," answered Young.

"Well, that airplane did not produce it, then?"

"That's right," answered Young. "At least, this report did not show it. Here again, let me state that this set-up was a very meager set-up compared to the later one which we made and with which we were able to investigate the possibilities much further."

"Where is the report of the last series of tests?" asked the Judge. "What was your recommendation?"

"The recommendation was 'None.'"

"'None,'" echoed the Judge. "Well, that cannot be very good, can it? No recommendation? You thought the plane was no good and you didn't want it."

"Let me offer this word of explanation," pleaded Young again. "The recommendation—if we had recommended further action, it would have to have been agreed to by the higher authorities, and with the present cut in funds by the Army Air Forces, we did not have sufficient funds to carry on with another project of much greater importance than this one, so this one had to be dropped. That is the reason we did not recommend any further action."

"But, if it had had merit," the Judge pressed, "would you yourself have made the recommendation 'None'?"

"If I was writing the report myself," Young responded, "and had been able to carry this through, I would probably have written a recommendation. But we agreed to make the recommendation 'None' and leave it neutral."

"That won't help me in judging if this thing had merit," snorted the Judge.

"Well, as I told you, the recommendation— if I had recommended that such and such be done—What was the purpose of it? I mean, we had no funds to operate on," repeated Young.

After this, Garvey asked Young a few more questions, and Reynolds threw a few weak jabs about fins, but Judge Curran seemed uninterested. He had come to a decision and wanted to finish the case.

When the last dribble of questioning had subsided, Judge Curran called the two attorneys to his bench.

He instructed them in low tones, both attorneys nodding, twice glancing at each other, then returning their gaze to the judge.

"Case dismissed," said the judge, sitting back and perfunctorily tapping the sound block with the gavel. He then rose and disappeared by the nearest door.

Willard, sensing that something significant had just transpired but not having a clue as to what it was, fixed a desperate gaze on Garvey, the word "dismissed" still echoing in his mind.

Garvey and Reynolds each returned to their respective tables, but Garvey beamed as he quickly stuffed his files into his worn leather briefcase.

"Over here," he said, leading Willard, Frank, and Don Young through a door into a small, adjacent briefing room. He switched on the light. "Close the door," he requested of no one in particular.

To Willard's puzzled expression, Garvey explained.

"You got your patent," he said through his smile.

Elated, Willard and Frank clasped hands in firm congratulations.

"But why did he say, 'Case dismissed'?" Young asked.

"You should understand better than anyone," Garvey answered. "The government is never 'wrong,' and so the judge won't actually rule against the government. Instead, he dismisses the case so that it appears the government wins, but it is with a stipulation.[15] In this case, the stipulation is: we revise the patent to state that the Channel Wing delivers vertical lift, not just superior lift, and then it will be a sufficient improvement over Henter, and the Patent Office will grant it."[16]

"And Reynolds agreed?" Willard asked. "He doesn't even believe the film!"

"He had to," Garvey answered. "The judge left him no choice."

"Well," Frank mused, "I would rather have heard him say, 'Decision for the Plaintiff!' or some such thing. It's very anti-climactic this way."

"This way everyone saves face," Garvey explained. "We still have to work with these people."

Unknown to Garvey, Frank had been working with some other people, including the Aviation Editor of *Popular Mechanics* magazine (Figure 3.5). He'd chosen that national periodical because of its circulation, and that month's issue appeared on the magazine racks just as the trial started. The feature article was about Willard's aircraft design. The front cover showed an artist's rendition of the Bumblebee flying above rural America. The title of the article had proved to be prophetic. It read, "The Wing that Fooled the Experts."

Nearly a year later on March 16, 1948, the United States Patent Office finally issued to Willard R. Custer, Patent Number 2,437,684 for "Aircraft Having High-Lift Wing Channels." The date was a full seven and one-half years after his application had been filed.[17] Fees for all the legal wrangling had cost the company more than $10,800.00.[18]

Figure 3.5 — May 1947 Popular Mechanics cover

IV The Experiment 1949 - 1953

My 'Chanfan' design and the limited model testing I pursued convinced me that airplane development while exciting, was only for the rich, with few customers.

- Thomas A. Hartman, Inventor

Now Custer reveals that he is working on another plane which "will fly this year before the snow does."

- "The Shape of Wings to Come?" *Flying*, September 1947[1]

"Ladies and gentlemen, it is my pleasure on this, the forty-sixth anniversary of the Wright Brothers' flight at Kitty Hawk, to present the Club's Wright Brothers Memorial Trophy to this year's recipient." The president of the Aero Club of Washington commanded the microphone. Hoisting the trophy to his eye level, he read, "'For significant public service of enduring value to aviation in the United States.'" He then turned to the audience, and said, "won't you please help me welcome — Mr. Charles Lindbergh!"

In the applauding audience that night of December 17, 1949, at the lavish Statler Hotel (today's Capital Hilton) in Washington, D.C., standing at round dinner tables just cleared of dessert, were scientists and aviation experts, all of them members of the Aero Club of Washington. Among them stood Willard Custer. Before Lindbergh's introduction, Willard had been networking with others at his table. He had just finished a long conversation with one of them, a lanky man about his same age: D. Barr Peat from Pittsburgh, Pennsylvania.[2]

In that conversation, Peat told Willard he loved aviation, even if he couldn't pilot a plane himself. In his late twenties he had created an aerodrome (a grass-covered landing strip) in his neighbor's cow pasture just outside the city. Barnstormers and local commercial clientele used the field heavily, and it eventually developed into the Pittsburgh-McKeesport Airport, later renamed Bettis Field. Peat had also been instrumental in establishing Pennsylvania's first airmail route, working with locals and Congress to get mail carried over the Allegheny Mountains by airplane from Cleveland, Ohio to Washington, D.C. with a stopover at Bettis Field. The first such flight had occurred on April 21, 1927. The area became an air traffic hub, and five years later traffic shifted one mile west to the Allegheny County Airport.[3]

As they talked at the Statler that evening, Willard learned that Peat had read several articles about the Channel Wing. Peat said he understood the concepts behind it. "You're the first," Willard smirked. The two arranged for Peat to visit Willard at his home in Hagerstown. That visit would last three days.

During those three days, Willard filled in the gaps for Peat, beginning with the barn roof, his early backing from Briggs Manufacturing, and Dr. Crook's inestimable help. He told him about the Bumblebee, Don Young and the Army Air Force testing, and the seven and one-half years of perseverance in which he had finally achieved his legal victory against the U.S. Patent Office and obtained suitable patent protection. He was awarded several follow-on patents on his idea.

His experiments continued, he explained to Peat. Don Young had documented in his Spring, 1946 test results that a shorter chord (front-to-back depth) in the channel gave superior performance to a longer one. He also had testified before Judge Curran that he believed that, given the right design and materials, an aircraft incorporating channel wings could be constructed that would take off without a runway.

"I'm putting those two opinions to the test," Willard said.

Willard showed Peat his second full-size Channel Wing aircraft, later named CCW-2 (Custer Channel Wing, Number Two). Willard had built this "test article," as he referred to it, in his own backyard. The objective: static, vertical lift.

Willard had handcrafted the channel wings and propellers himself (Figure 4.1).[4] The channels were six feet in diameter, the same size as those on the Bumblebee. But the chord length on the channels (front to back) measured only 32 inches, much shallower than the Bumblebee's six feet,

Figure 4.1 — Willard, Dr.Crook, and Frank discuss CCW-2 construction

to test Don Young's recommendation. Mounted inside each channel hung a 75-hp Continental engine, with the propeller in the back, pusher-style. Willard had mounted a wheel below each channel for landing gear. Each channel, configured in this way, he then attached temporarily to a frame of two-by-fours that held it upright and provided for each channel to be tested independently. The frame included a hinge on the ground that would allow the channel to move up and down in the frame. With Curley, Willard's oldest son, now a civilian, on one end of the channel, and another helper on the other end, Willard had tested each channel by simply revving the engine. The channel and propeller generated enough static lift to raise the channel, engine, wheel and two helpers three feet off the ground. At least two journalists witnessed it, and at least one photographer captured it (Figure 4.2).[5]

Satisfied with their lifting capability, he then attached these signature features to a salvaged Piper Cub fuselage and tail. Piper Cubs enjoyed wide popularity, but on this craft the lineage of the fuselage was almost unrecognizable. The craft had been rendered as lightweight as possible, reduced to a simple tubular steel frame with no skin and without the Cub's typical nose engine. It had no wings of the conventional variety, no enclosure around the

Figure 4.2 — Willard, Curley and helper demonstrate vertical lift of a CCW-2 channel wing

cockpit for the pilot, and minimal instrumentation. Except for the twin engines and three wheels, the aircraft appeared to have been stripped for parts. Without the pilot, it weighed less than 1,000 pounds[6] (Figure 4.3).[7]

Officially, the CCW-2's maiden flight had taken place on July 3, 1948, at the Hagerstown Municipal Airport, piloted by Willard's son, Curley, by then a seasoned pilot, recently separated from the Air Force. (Unofficially, Frank Kelley had also taken it fifteen feet into the air quite accidently during a "taxi test" prior to that.) The aircraft had no brakes, flaps, conventional wing area, or even ailerons, so it could only be flown in a straight line.

"What do you ultimately want to do?" Peat asked.

Willard replied, "We want to design a light plane that will land and take off without a runway. It will cruise at more than 100 miles an hour. It will be as controllable as any other airplane and have the same equipment. Only difference is, you'll keep your brakes on until you're practically airborne. Then you'll release the brakes—and you will be airborne.

"Landing, it will be the same thing in reverse. You'll come practically straight down. When your wheels touch, jam on the brakes. Fifteen feet later you'll be stopped.

"If the engine fails, you will be able to bring it in like any other conventional model, although it will land faster. But you'll still be able to slow it down quickly with brakes. And even if you should crack it up, it will not be a

Figure 4.3 — Frank and Willard pose with the CCW-2

major financial disaster. I figure a good manufacturer can stamp out channel wings in wood or metal like hot cakes. Even including engines, a two-placer should cost no more than $700."[8]

By the end of his visit, Peat concluded that Willard had gotten what Peat called "the Billy Mitchell treatment" by the U.S. Army Air Force. Willard remembered the story. Billy Mitchell had finished WWI in charge of all U.S. air operations in Europe. After the war he insisted that an airplane could sink an enemy ship better than another ship could. His superiors disagreed with him and sidelined him. He later publicly accused both the Navy and War Departments of "stupidity" and "incompetency, criminal negligence and almost treasonable administration of the national defense." He was promptly court-martialed for insubordination and quit the service. After his death in 1936, however, his views on airpower—if not his opinions of his superiors—were vindicated. Posthumous medals and memorials followed, and the Mitchell B-25 bomber was named for him. Peat believed that Willard, like Mitchell before him, had presented the military with the vision of aviation's future, but the military leadership had again proved itself incompetent in its decision to pass on developing the Channel Wing.

So, Peat promised to try to help. And when he returned to Pittsburgh, he went directly to Jack Heinz.

H. J. Heinz II ("Jack") was the grandson of the H. J. Heinz who had established Pittsburgh's H. J. Heinz Company, a food company today universally recognized primarily for its ketchup, but the producer of Smart Ones®, OreIda®, and Classico® product lines, as well. Educated at Yale and Cambridge, Jack had inherited the Heinz dynasty and management of the company and expanded its reach into international markets. Heinz' ex-wife had been an aviatrix, so he knew the enthusiasm the public had for flying, which is to say, he perceived the opportunities to be had in the growing aviation market. A handsome member of the rich upper class, he enjoyed his privileges, but he was also a genuine philanthropist, serving at times as chairman of the United War Fund during WWII, the Community Chest (which became the United Way), and the Howard Heinz Endowment by which he helped rejuvenate Pittsburgh's downtown.[9]

Peat's new find intrigued Heinz, and Peat arranged for Willard to bring his CCW-2 to the Allegheny County Airport where Heinz and others could see it fly.

Test flights proceeded in earnest throughout July 1950, and again the following summer. Curley logged over 300 hours in short flights around the airports, both in Pittsburgh and Hagerstown.[10] Why so many hours? It felt like flying with the top down and windows open.

"It was fun," he said, smiling boyishly (Figure 4.4).[11]

Willard recorded that the plane could take off at a mere 30 miles per hour after a short 45-foot roll, and the highest static lift measurement he had taken was 1,144 pounds.[12]

The CCW-2 flights were also excellent opportunities for Barr Peat, who continued to work his contacts in and around Pittsburgh. Some invitees became interested in the unique aircraft itself, while others were interested in the investment opportunity.

One person interested in the aircraft was C. Gilbert ("C.G.") Taylor, who knew a great deal about designing and building airplanes. Born the same year as Willard in Rochester, New York, he had taught himself how to design light aircraft. In 1927, he and his brother Gordon had started the Taylor Brothers Aircraft Corporation with an aircraft of their own design, the first ever side-by-side two-seater aircraft they aptly called the "Chummy." They were both pilots and they demonstrated the Chummy in many barnstorming exhibitions across the country. Tragically, while flying with a passenger

Figure 4.4 — Curley flies the CCW-2

in one such demonstration in Detroit in 1928, Gordon crashed and was killed.

Though his brother's death drained his tolerance of risk, C.G. persevered. He eventually moved his family and the business a few miles south to Bradford, Pennsylvania on an invitation from that town's Chamber of Commerce. There, his company designed and produced a glider with two seats, one in front of the other (in tandem), as well as a single small-engine aircraft he called the Taylor E-2 Cub. It featured an open cockpit, a tubular steel fuselage covered with cotton fabric, and wooden wings.

During the Depression, the company filed bankruptcy but re-emerged under the control of William Piper, a Pennsylvania oilman of considerable financial means and business savvy. Taylor still held the position of company president, but Piper remained the secretary-treasurer, and in a power struggle that ensued Taylor was bested. He sold his interest to Piper and departed the company in 1936. Piper continued to develop Taylor's E-2 Cub, modifying it slightly to become the J-2 Piper Cub under the company's new name. Piper eventually sold more than 20,000 of the little planes. Willard had built the CCW-2 on a frame of one of them.

Because of Taylor's reputation, the Chamber of Commerce of Alliance, Ohio invited him there to begin again, and so there he set up the Taylorcraft Aviation Corporation. He immediately revived his Chummy design in the BC-12 series. By 1939 Taylor had resumed producing eight of the airplanes every day and employed 750 people. Then tragedy struck again. A fire broke out in the company hangar, destroying many planes, and—most upsetting to Taylor, a devout Jehovah's Witness of uncommon compassion—it took the life of a young man who had run back into the burning hangar. Lacking the stomach for the internal company power struggles that followed, Taylor again exited his own company in 1942.

By 1947, he had bought what pieces remained of the bankrupted Taylorcraft Aviation's assets, and with them he started his third company, Taylorcraft, Inc., "Builders of America's Favorite Light Planes for Business and Pleasure."[13] That year he also saw the CCW-2 and typed a letter to Willard.

> I would like very much to drop everything right now and go to Hagerstown and work with you on this. Unfortunately I have a big problem on my hands and that is to find a place for this Company to operate after our present lease expires at the end of October. I will keep working on it, however, in my spare time and the first opportunity I get after settling the location, I will be in to see you.

He settled in Conway, Pennsylvania, just ten miles up the river from Pittsburgh.

True to his word, C.G. returned to Willard with interest, convinced that Willard "had the answer to the problem of lift without velocity in light aircraft," in Frank Kelley's words. C.G. proposed that he license the Channel Wing concept from Willard and incorporate it in a few aircraft designs that he would develop and manufacture through Taylorcraft.[14]

Licensing the Channel Wing concept constituted Willard's preferred business plan. "It is the intention of the corporation to license manufacturers to use the Custer Channel Wing in aircraft of their own design," he had said in his company's publication. Willard preferred to "continue research and development on other uses of the wing rather than get into actual production."

Unlike C.G. Taylor, Willard did not have the wherewithal to manufacture aircraft. He had no money beyond what people gave him. Nor did he have the machinery, employees, warehouse, manufacturing plant, management experience, business acumen, airline industry savvy, professional network, or business education to build and sell aircraft successfully. If he could license his wing to others, such as Taylor, with the background and resources to conduct a manufacturing program, then Willard could avoid having to overcome his own handicaps in that regard.

"I'm a mechanic," he said to Flying magazine. "I want to keep on tinkering and let somebody who knows about it worry over assembly lines." It appeared to be a match made in heaven, so Willard and C.G. seriously discussed the possibilities.

C.G. liked the idea of a commercial two-seater, like his Chummy, with a channel wing. Willard still had the idea of building a military liaison plane, as the Army Air Forces had previously considered developing. It could carry a pilot and two patients with a medical attendant, and since the war in Korea was churning out a lot of patients, this could fill a sad need. Because the Army had previously considered it, C.G. agreed to at least start with that concept, and the two aircraft designers struck an agreement and prepared to announce it to the world.[15]

Heinz became more interested, too, but he had a different idea and a different bank account. He offered to finance what he considered to be Willard's next step, a commercial aircraft that could be sold to people like him, corporate executives who wanted personal planes that could land on private landing pads away from busy airports. Heinz, in fact, had already contacted Jack Baumann about the idea. Baumann had been an employee of Taylor's in the pre-Piper days. Since then, he'd moved on to work for Lockheed, and then started his own company, Baumann Aircraft Company, in Pacoima, a suburb of Los Angeles, California. Baumann had been in the news recently with a prototype of his own, the Brigadier B-290, a twin-engine, five-seat, all-metal executive transport airplane.[16] The aircraft sported a distinctive feature that grabbed Heinz' attention: its rear-mounted (pusher) propellers.

So, in 1950, Willard—directed and funded by Jack Heinz—traveled to Los Angeles to meet with Baumann, the president and chief engineer of his own aircraft company. Heinz said he wanted a commercial prototype, a finished product that provided a concrete example of channel wings in commercial use. Willard explained how to build and integrate the channels into Baumann's design. Heinz agreed to pay Baumann a hefty price and took title to the aircraft. Then Baumann set to work on what would be named the CCW-5, the Custer Channel Wing five-seater.

On August 29, 1950, under Heinz' influence, Willard's company got a much-needed corporate makeover. The Custer Channel Wing Corporation replaced the National Aircraft Corporation with Willard as the company president. Frank would stay involved in the background.

The company installed Willard's daughter Helen as the full-time, salaried company secretary and Willard's administrative assistant. She ordered stationery with a corporate logo, the Hagerstown address, and all those other trappings which herald a company's transformation from a backyard avocation into a bona fide corporation.

Helen, now twenty-nine, had already grown to become Willard's right hand. She had her mother's diminutive build and tiny waist, but she had her father's energy, a low voice, and coal black eyes that twinkled behind black cat-eye frames. She had been typing all of Willard's correspondence, personal and otherwise. She now became his office manager, opening all the mail, controlling his calendar, arranging his transportation, taking dictation, publishing meeting minutes. She was privy to every corporate secret. She even became a notary public so she could notarize any necessary papers for the company. But her best traits were her loyalty (Willard was still "Daddy"), her genuine sweetness, and her political savvy.[17]

"Frank, it's C.G. Taylor. I've got three memos here that say I owe you a call back."

"That's okay, C.G. I know you're out on the factory floor, keeping the machines running."

"That's a fact, sir. How's my friend, Willard?"

"Willard is just back from Los Angeles."

"Los Angeles? What, is he starring in a movie?"

"No, C.G.," Frank laughed. "Mr. Heinz has contracted with a fellow, Jack Baumann, to put channel wings on a commercial-style plane of his, and he sent Willard out to go over the details."

"Jack Baumann, you say. I worked with a Jack Baumann once, back before the Piper days. I thought he'd left to work for Lockheed."

"Well," said Frank, "it's a small world. He runs his own company now. What did you think of him when he worked with you?"

"He's a good man," C.G. recommended. "Willard will do all right by him. So, Baumann's going to build a fleet of these and start selling them?"

"No," Frank answered. "Just the one. It'll be a prototype, an example of a commercial Channel Wing that Willard can show."

"Showing is one thing; selling is another. I guess if you're in the showing business, Hollywood is the place to be."

"That's not Willard," Frank replied. "If he's showing something, it's with the aim of selling it."

"Heinz needs to build him a fleet, then, not a prototype," C.G. spat.

"Which brings me to why I called you," Frank said. "We're still focused on Taylorcraft building a fleet of Channel Wing liaisons. How are the prospects looking?"

"I'm still priming the pump, Frank. I think there's water yet in this well but getting her flowing again will take some time. I can't do anything with the Channel Wing until I get this factory going again."

"I understand, C.G. We want to do a demo with you in Pittsburgh. We figure showing a 1,000-pound CCW-2 that generates 1,100 pounds in static lift will get us some headlines. But we have to time it right for you. If we get a flood of phone calls, we need to be ready to close deals."

"My crystal ball is foggy, Frank. The war in Korea has grabbed up all the pilots, and there's not much interest in buying commercial planes right now; that's the word I'm hearing. Willard is probably right about the military; that might be his best strategy."

Frank replied, "Yes, we will definitely go back to Gen. Gilmore and see if there's an opportunity there. Willard is planting seeds all over, ready to go wherever the crop grows."

"That's what you gotta do," C.G. agreed.

"So," Frank returned, "how about we give you the rest of the year? Can we demo, say, in December? Maybe Santa will bring us a happy new year to follow."

"Sure," C.G. acquiesced. "But you know I don't believe in Santy Claus."[18]

Midday Thursday, December 6, 1951, two black sedans, one towing a trailer, converged on the windsock pole at the Allegheny County Airport, four miles southeast of Pittsburgh. Lower than the nearby Lebanon Church Cemetery, the airport barely fit on the knoll, the end of its secondary runway dropping precipitously into the Monongahela River below. A small crowd had gathered, including Barr Peat, his family, and several newsmen. Cloudy skies and still air allowed the temperature to reach 62 degrees Fahrenheit.[19]

Willard and three other men stepped out of the sedans. Forty-year-old Walter ("Walt") J. Davidson, a pilot and a dealer for Texas Engineering and Manufacturing Company (TEMCO), selling Swift 125 aircraft at this airport, had joined the team. Walt, C.G. and another man easily off-loaded the CCW-2 down a ramp from the trailer, maneuvered it to the windsock pole, and tied it down. A steel cable extended from above the cockpit and bound the airplane to the pole, its tail pointed inward toward the pole and its nose faced the small audience of invitees. Another cable about five feet long anchored one of the wheels to a twelve-inch metal rod set in the ground. This plane was going nowhere.

Peat greeted his prospects while Frank checked in with the journalists.

Each man took his station. C.G. and the other man gripped the stubby wing tips on the outside of the channels. Walt and Willard stood on opposite sides of the plane's empty cockpit. Checking that all were ready, Willard reached inside and started the engines. The propellers spun into a blur. Mounted on the backside of the channels, they posed no danger to Willard or Walt. The CCW-2, eager to fly, strained forward against the two cables, stretching them taut. Still standing confidently by the cockpit, his fingers on the controls, Willard slowly increased the gas flow to the engines.

Little by little the engines grew louder and the propellers buzzed faster. Then, as if coaxing a baby bird from its nest, Willard tapped the throttle just past that magic threshold where lift overcomes gravity, and the aircraft's wheels released the ground, floating as high as Willard's reach would allow. Willard remained by the plane, his hand barely reaching the controls now, as the plane, light as air and steadied by a man at each wing tip, responded to his touch. It felt wonderful, and he recalled his first model, rising vertically from the tabletop, restrained only by strings tied to his fingers. Unable to hear anything over the roar of the engines, Willard glanced about and, spying the camera man, nodded to him to take the picture.

The camera snapped a scene for all to see, for all time: a full-size aircraft, with self-produced "speed of air" coursing through its channel wings, lifted by air pressure from a complete standstill—like a roof from a barn in a wind storm—hovering against the insistence of rooted cables. Also, pictured—for any persons who might appreciate it—at the top of the pole a deflated windsock hung limp. The propellers alone had provided the moving air required for lift; nature had provided no supplemental breezes (Figure 4.5).[20]

News of the demonstration rippled through the news media. A reporter from the Pittsburgh Press proudly reported the story and announced the company's agreement with Taylorcraft, Inc. in that evening's circular. The next day, the same newspaper ran a picture with a lengthy caption. A week later TIME magazine (Dec. 17, 1951) reported the event. A month later, the London, UK periodical The Aeroplane (January 1952) recounted it and its

Figure 4.5 — Walt Davidson and Willard demonstrate the CCW-2

associated announcement:

> During the demonstration an announcement was made that Taylorcraft Incorporated, of Conway, Pennsylvania, had been licensed to build the Custer Channel aircraft for commercial and military use...[A]ccording to W.R. Custer, ...it is the intention of his company to license the industry and not to manufacture, unless forced to do so...[T]he first type to be projected with the Custer Channel Wing by Taylorcraft Inc. on an experimental basis will be a military aircraft that will comply with the U.S.A.F. specification for a liaison aircraft.

In its final demonstration, this aircraft had defied some staunchly held aeronautical theories and assumptions, and many aviation experts choked on its exploit. For example, David Anderton, a writer for *Aviation Week* who attended the demonstration, remained skeptical. In the December 17 issue, he confessed, “This writer...is not convinced he saw vertical lift,” suspecting some sort of subterfuge with the cabling.

Planting Seeds

Shortly afterward, in February 1952, Dr. Hugh O'Neill, a professor of botany and a colleague of Dr. Crook at Catholic University, wrote Willard to introduce an idea that resounded with Frank Kelley.

> Enclosed are ten copies of our recent conference to discuss my idea of using your channel wing plane as a means of a new type of aerial photography which should all but revolutionize military reconnaissance photography...It seems very evident to me that your plane offers better facilities for taking aerial photographs for purposes of forestry, game management, land management than any other plane now available. Mr. William Negel's idea that your plane could offer very special advantages for fighting forest fires, ships at sea on fire, etc. could be capitalized immediately. I hope you will call us on your next visit to Washington.[21]

Then, Willard called Gen. Gilmore at Wright Field. He might be interested to see his liaison aircraft when he prototyped it. Maybe the Air Force would consider a Channel Wing for its reconnaissance planes? Would he like to run tests on the CCW-2? No, Wright Field had no opportunities at the moment.

Instead, the general redirected Willard to the National Advisory Committee for Aeronautics (NACA). With financial resources beyond that of private industry, Congress had created NACA in 1915 to research and advise (as its name suggests) the federal government (directly) and the U.S. aviation industry (indirectly) on matters regarding aeronautics, including policy and technical solutions. To enable its research, NACA constructed several special-purpose wind tunnels at the Langley Memorial Aeronautical Laboratory, now Langley Air Force Base, in Newport News, VA. Seven years later, on October 1, 1958, NACA would be folded into the National Aeronautics and Space Administration—NASA.

In July 1948, the month the CCW-2 had made its maiden flight in Hagerstown, a NACA engineer visited to see it and collect some data for a report. NACA told Willard it was interested in "airplanes capable of hovering and flying at very low speeds" without sacrificing cruise speed, so, they would be interested in studying the CCW-2. By October 1951, Willard had concluded his experiments with the CCW-2. Confident he did not need to test it further, he canceled its CAA registration, retiring it. Late in 1952, then, Willard loaned the CCW-2 to NACA to run a series of tests in their full-sized wind tunnel at the Langley lab; but the tests would run without his input or participation.[22]

With the CCW-2 recently in the headlines, the Custer Channel Wing Corporation's telephone came alive. To respond to inquiries, Willard developed a concept paper, published March 1, 1953, to describe what he envisioned a Channel Wing aircraft could do:

> The Custer Channel Wing is a fundamentally new concept in aircraft design which will permit a ship equipped with the Custer Channel Wing to:
>
> 1. Take off in a few feet or rise vertically and hover if properly designed for such performance.
> 2. Attain forward speeds in excess of those of conventional aircraft.
> 3. Slow down, hover in midair and land vertically under full control.
> 4. Substantially increase payloads, flight range and endurance over that of any other ship.

The claims were visionary. No Channel Wing aircraft had yet been built that could deliver these performance targets. The caveat to all of the claims was expressed most succinctly in the first one: "if properly designed for such performance." The same caveat applied to all of them as the body of the paper explains. Nevertheless, the paper expounds on each claim with soaring overstatements.

For example, about power-induced lift, the paper says, "Custer Channel Wing in full scale has already demonstrated statistically an infinite lift coefficient. This compares with the conventional plane's average lift coefficient of two, which in some instances has been increased to five by adding flaps and powered lift devices." The contrast between the Channel Wing's "infinite" and the conventional plane's "two or maybe five" would certainly mark the Channel Wing as superior except that "infinite" is unbelievable.

Again, regarding the Channel Wing's forward speed potential, the paper says, "[A] ship with channel wings can attain forward speeds impossible for a helicopter and in excess of those obtained by any fixed wing aircraft." This tone resonates with that of the fourth statement above that the Channel Wing could essentially out-fly any other airplane. No doubt these claims were mostly sales hyperbole designed to inspire new thinking, but they may also have hurt the aircraft "in terms of engineering credibility," as aviation historian, Walt Boyne, reported.

One superlative that may not have exaggerated very far appeared in the

"Application" section of the white paper. "There are [sic] an indefinite number of adaptations of the Custer Channel Wing. It is adaptable to any type of ship from the small private plane to large transports." As examples and illustrations, the paper presents pictures of models and artist's conceptions of a range of ideas, from one-channel flying cars to supersonic jets that look like missiles.

Willard intended to inspire innovators and geniuses with this paper. The Custer Channel Wing Corporation readied for an incoming tide of inquiries from companies looking to take advantage of the opportunity to design and produce such aircraft. Willard and his directors expected the paper to kindle a revolution throughout the aviation industry.

NACA Research Memorandum L53A09

The very next month, April 7, NACA published its findings from its study of the CCW-2, officially titled, "Research Memorandum RM L53A09, Langley Full-Scale-Tunnel Tests of the Custer Channel Wing Airplane" by Jerome Pasamanick. The report did not echo the superlatives of Willard's concept paper. As he read it, Willard wished he could have been involved in the testing with Pasamanick as he had been with Don Young.

The NACA report is very technical, but a summary chart shows that, with the propeller rotating at 2,625 revolutions per minute, Pasamanick measured the maximum static lift to be 904 pounds. He estimated that the plane itself weighed about 900 pounds. Yet he concluded, "The controls were completely inadequate under static lift conditions and the airplane could not be flown in hovering flight." Willard, of course, had not designed the CCW-2 to be a hovercraft and so had no controls for hovering flight. The only controls employed during static lift demonstrations were steel cables tied to immovable objects and Willard's arm reaching in from outside the cockpit.[23]

The NACA report did, however, contribute important findings from its study of the airflows around the propeller positioned at the trailing edge of the channel on a stationary aircraft.

For example, Pasamanick noted that, as the propeller pushes air behind itself, some of that air recycles around the tip and back to the front of the propeller. In his diagram (Figure 4.6), this airflow is indicated by an arrow snaking down from the top left corner. He noted that this immediate recirculation occurs only in the upper half of the circumference circumscribed by the tip of the rotating propeller, that is, above the channel. The channel blocks this immediate return in the lower half of the circumference. What

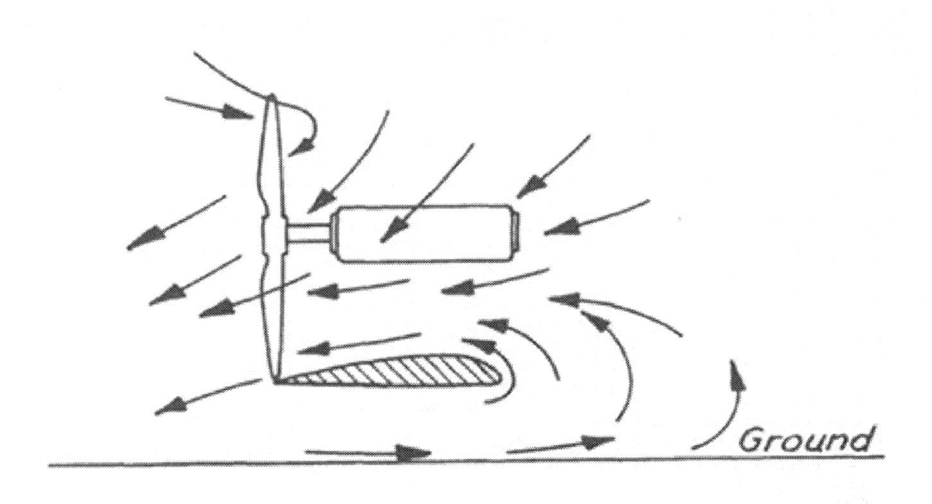

Figure 4.6 — Air flows around the propeller (Pasamanick)

Pasamanick did not note, and so perhaps did not appreciate, is that the channel's blocking this immediate air return helps to further lower the air pressure within the channel and thereby generates even more lift. Willard understood this and had described it in his letter to the US Patent Office in June 1941, ten years prior. In fact, Willard had found that, to block this return airflow most effectively, the propeller had to be placed as close as possible to the back edge of the channel.[24] But some propellers already on the market, when rotating, would bend forward at the tips so much that they would shave the back of the channel, damaging both wing and propeller. Willard had fashioned the CCW-2 propellers himself because he could not find a suitable pair on the market.

Another important finding of Pasamanick's study documented the decided downward angle of airflow behind the propeller. While the air stream coming off the back of any airplane wing typically flows slightly downward, "A visual tuft survey in the region behind the propeller ... showed that the propeller slipstream for the static flight condition was deflected well downward and underneath the tail." He describes it again as an "acute downward deflection" and notes that the air stream passed "completely below the tail." This deflection would become an issue.

The entire tone of the NACA test report rang of failure. The report was too technical for the public to digest and critique. But the important fact is that it failed to provide the hoped-for recommendation to the industry to adopt the Channel Wing design. Had the report been more optimistic, the industry might have formed an altogether different opinion of the radically different technology. As it happened, however, the report set Willard once again in opposition to the federal government's aviation experts, a familiar but disheartening place to be. Willard was experiencing another "Billy Mitchell" moment, as Barr Peat might say.

Just a month before, the Custer Corporation's white paper had made some lofty projections of a Channel Wing's capabilities. But after NACA published its report, Anderton—the Aviation Week writer who'd said he didn't believe the CCW-2 had demonstrated vertical lift—penned a follow-up article entitled, "How Good Is the Custer Channel Wing?" In it he wielded the results of the NACA report like a machete to hack down every claim the Corporation had published in its concept paper. Anderton concluded: "This all adds up to almost complete deflation of the claims made by Custer."[25]

When asked how he remained positive after suffering such disappointments, Willard responded with his characteristic sense of self. "At first I just got mad," Willard admitted. "Then I realized that I was just too far ahead of my time and went back to work."[26]

V The Prototype 1953 - 1957

This is a new breed of aircraft. We're going to build aircraft that will usher in the air age. We haven't had the air age yet because most of the aircraft have the high lift taken out of 'em trying to go to the moon. We can go to the moon, but we don't know how to fly 10 miles an hour.[1]

- Willard Custer

The starting gun would not fire until October 8, 1953, but earlier in the year, entrants were lining up for the first and only London to Christchurch Air Race. The course, extending 12,365 miles, began at Heathrow Airport and followed a direct line southeast across Europe, Iran, India, and Indonesia, continuing the whole of Australia, then terminating in Christchurch, New Zealand. Most of the entrants would be flying British aircraft, but some Dutch aircraft were also entered. The KLM Royal Dutch Airlines Liftmaster, for example, took its place on the roster, carrying fifty Dutch immigrants, twenty-six of them brides-to-be, anxious to marry on arrival in New Zealand. Observers favored this plane to win the race (and it did win the transport handicap category) for reasons other than the energy generated by the anticipation and urgency of its passengers.[2]

The British publication, *Flight*, reported on February 27, 1953 that the Americans had placed an entry: the Custer Channel Wing, CCW-5, still under construction in Baumann's production facility in Ventura County, California per the contract with Heinz. The reporter described the aircraft as "easily the most unconventional entry" and "almost completed." He also remarked wryly, "Whether this implies that it will be ready to race half way [sic] round the world by next October is, perhaps, another matter."

The British periodical optimistically assessed the Channel Wing's chances. "The true potentialities of the CCW-5 are unknown," it said in February, "and the type could prove something of a dark horse." The article mentions the plane would be using two Continental SO-470-B engines, which provided at least 280 horsepower at 3,000 revolutions per minute and posting a cruising speed of 230 miles per hour. The target speed of the transport handicap, it pointed out, "is no more than 195 mph., which augurs well for the Custer's chances." Still, it cautioned in general, "some aircraft, particularly the smaller types, will be forced to put down for refueling with distressing frequency." Of course, the magazine could only guess because the CCW-5 had never even flown.

The Channel Wing joined the KLM Liftmaster in the transport handicap category. One of the British entries in the same category would be carrying no fewer than three flight crews to maintain crew alertness over the duration of the sixty-hour trip. But the CCW-5, a five-seater, could accommodate a full crew of no more than five members. The American team offered the name of only one member of its crew, specifically, Walter J. Davidson, the pilot. He may have thought he could fly the race solo.

Walt Davidson had met Willard in Pittsburgh the previous year. He worked in sales at the Allegheny airport when Willard and his son Curley were demonstrating the CCW-2 there. He quickly became enthused with the aircraft and helped Willard conduct his tethered static lift experiments. With the CCW-5 nearing completion in California, Walt signed on to be Willard's primary pilot, permanently moving his Canadian wife, Estelle, and their two children to Oxnard, California, just blocks from the airport where the CCW-5 would be hangared.

Walt brought an impressive resume.

Born in Pittsburgh, he had left for Montreal while in his late twenties, as the German Luftwaffe bombed London, post-Dunkirk. The U.S. had not yet entered the Second World War, but Roosevelt agreed to supply Churchill with bombers and ordinance to fight back. Someone had to secretly deliver the U.S. bombers to London, and England's less hesitant North American ally—Canada—recruited hundreds of crack Canadian and U.S. pilots for the job. Walt had been one of them.

Walt discovered that flying brand new, untested, Lockheed Hudson bombers 3,785 miles from Gander Airfield in Newfoundland to Prestwick in England felt dangerous but thrilling. To maintain secrecy, the flights were all made at night under radio silence with no radar, with only the black abyss of the Atlantic below and the guiding stars above, should they be visible. Each bomber had two engines but no sound insulation, no heat, no pressurized cabins, and a fuselage overladen with extra fuel tanks, as the planes weren't designed to fly that distance without refueling.[3]

That was only the beginning. By the end of the war, Walt had also flown a B-24 Liberator 11,500 miles to Cairo, Egypt via Brazil and Ascension Island, and a B-25 Mitchell to Australia by island-hopping across the Pacific. This "Ferry Command" experience had emboldened him, now 43 years old.[4] The notion of flying a transnational, 12,000-mile, sixty-hour air race all by himself in a newly assembled and untested airplane packed with extra fuel tanks sat comfortably with him. He had a friend in New Zealand from his Ferry Command days, Ernle Clark, described by his hometown as "the second person to ever fly solo from England to New Zealand."[5] If Ernle could do it, maybe he could, too. Walt wrote Ernle with bravado that he was "looking forward to winning [the race] and thus properly advertise this very important Aeronautical Discovery [the Channel Wing] to the world." Like Willard, Walt valued the free advertising to be had in a dramatic, front-page newspaper story should he win.

By July, Baumann had readied the plane for its maiden flight and hangared it at the Ventura County Airport in Oxnard. On Walt's first visit to the hangar, Willard introduced him to the aircraft.

Walt approached the new airplane slowly, almost reverently. Painted primarily in high gloss white with a red cowl atop the nose and cabin, it resembled nothing he'd ever seen. Walt, like everyone who first sees a Channel Wing in real life, found his eyes riveted on the wings (Figure 5.1).[6]

He approached from the passenger side, in front of the wing, then stopped, taking in the sweep of the seven-foot channel, noting how it emerged from the fuselage, then swooped under the suspended engine to re-emerge, level off and taper into a conventional wing tip.

He rounded the end of the wing, finding the black propeller attached to the rear of the engine, one blade measuring the channel's radius, the other standing upright like a sentinel, overlooking the nacelle, which housed the engine.

As he rounded the rear, he studied the tail, impressing its vertical and horizontal stabilizers into the soft clay of his memory (Figure 5.2). Walt mapped every inch of the plane into his mind's eye, as a blind person surveys a room with his hands and feet to memorize and measure each obstacle, each feature, for future navigation. Similarly, Walt imaged the plane in

Figure 5.1 — CCW-5 Front

his memory, preparing for the moment when he'd be in the cockpit, unable to see much behind him, but able to picture it from memory.

From the tail, Walt admired the curvature of the two wings. A Channel Wing plane is most interesting when viewed from the rear, he noticed. Here, the intersection of the wing with the fuselage is the most graceful. The interplay of channel and propeller can only be observed here.

Finally, Walt approached the nose. He touched it. His fingertips caressed the aluminum, tracing the fuselage to the pilot's window, then to the juncture with the channel. Then he turned, as if released from his trance, looking for Willard.

Willard still stood near the door of the hangar by which they had entered. He had watched his new pilot's first meeting with this extraordinary aircraft.

"Is it open?" Walt called.

Willard nodded.

Walt turned the handle of the cabin door, lifted his weight and slipped

Figure 5.2 — CCW-5 Rear

into the snug cabin. His eyes followed his fingertips as he learned the arrangement of the instruments, touching switches, noting labels, caressing the yoke, positioning his feet on the pedals, trying to see around his knees. As if cinching strap-on wings, Walt tested the fit of this new plane. The channels, the tail, the cables and switches would all become extensions of his hands and feet; like prosthetics, they would enable him to do things his natural limbs could not. He would have to practice long in the air until each gust of wind, each change in orientation would, without thought, trigger his muscle memory to take the proper action to make this ship do instantaneously what he wanted her to do.

On July 13, 1953, three months before the race, Walt piloted the CCW-5 aloft for the first time (Figure 5.3).[7] He tested it in four hops at the little Ventura County Airport. In each hop, Walt lifted off at forty miles per hour after running 250 feet down the runway. Four days later, he shortened the takeoff run to only ninety feet at thirty miles per hour, a feat extraordinary enough to be reported in *TIME* magazine. The plane exceeded his expectations in all but one aspect. Instead of the SO-470-B engines (280-300 horsepower) that had been specified, the Continental company could only supply O-470-A

Figure 5.3 — CCW-5 lifts off

engines at 225 horsepower each. This Channel Wing depended on high horsepower engines and proper propellers to achieve optimal performance, so Walt suspected that its STOL capabilities were suffering. He had two months to find out.

Through August, Walt pushed the CCW-5's performance levels, consulting daily with Willard back in Hagerstown. What technique best shortened the takeoff, the landing? How slow could it fly? When, if ever, would it stall?

By September 10, a month before the race, he had his answers. Walt wrote to his Ferry Command friend, Ernle Clark, in New Zealand, with the bad news that the "Continental Engine Co. has been unable to supply us with the approved type superchargers and during recent tests on the aircraft I found that the present propellers are not satisfactory." It would take at least six more months to rectify the technical problems, he wrote, and that effort superseded all others. Disappointed, Walt quietly withdrew the Channel Wing from the race. But not a bit discouraged, he pitched to Ernle, "NOW is the opportunity to get in on the 'ground floor' of something new in Aviation. All future aircraft will HaveTo [sic] adopt this wing to stay in production." Maybe Ernle could get something started with the design in New Zealand, he suggested.

The Type Certificate

So, Willard had a real Channel Wing airplane, not a testbed or an experiment. The CCW-5 embodied one possible commercial application of

Channel Wing technology. But it had one critical imperfection unrelated to technology: Willard didn't own it. Jack Heinz had paid for it and held its title. And Heinz had clear expectations. He expected that the CCW-5 would be the prototype for a line of executive aircraft to shuttle businessmen between corporate offices. He expected Willard, with his corporation, to get it certified by the CAA for production. To accomplish this, Heinz also expected the Custer Channel Wing Corporation to make the leap from small-town, backyard operation to a bona fide aircraft manufacturing business with assembly plants and engineering and sales departments.

Willard's view was different. He had the use of the CCW-5, and he had a totally devoted, expert pilot in Walt Davidson. On the heels of the disappointing NACA report by Pasamanick, Willard could now show the aviation industry and the military what a Channel Wing could do and the advantages inherent in it.

Further, Willard did not know how to grow his quaint, neighborhood corporation into a large manufacturing business, as Heinz envisioned. His plan had always been to commercially license the Channel Wing idea to any number of existing aircraft manufacturers. Under this arrangement, a

Figure 5.4— Pilot Walt Davidson[8]

commercial firm would pay Willard an annual licensing fee for the right to design, develop and manufacture aircraft incorporating a channel wing. The licensee would assume all the risk and costs of inserting that aircraft into the commercial aircraft industry, and Willard could watch from the sidelines and count his money. One such manufacturer, Taylorcraft, Inc. in Conway, Pennsylvania owned by C.G. Taylor, the inventor of the Chummy and the Cub, had agreed to manufacture two aircraft models with channel wings. So far, however, C.G. was still working to realize a profit on his conventional aircraft so he could reinvest it into his first Channel Wing model, a liaison type plane for the military. Consequently, he was not actively licensing anything nor paying any licensing fees, as of yet.

Speaking of the military, there was another option. Ever since the Bumblebee testing at Wright Air Force Base, landing a military contract had been an alternate goal. A realized contract might resolve all of Willard's financial concerns. With it he might negotiate an ample annual licensing fee, and for a man living on $100 per week, the fee need not be exorbitant to be ample. He could watch from the sidelines as his Channel Wing revolutionized air travel, and it wouldn't cost him another nickel.

Right now, Willard reasoned, the CCW-5 needed to perform well. Then the military would offer a contract, or other established commercial aircraft manufacturers would beat down the door, with cash gripped in their fists, competing to be the next company to be licensed to build Channel Wing aircraft, little ones, that the common man could fly. Until those offers materialized, however, Willard had no choice but to follow Heinz' plan. He and Heinz both agreed that this Channel Wing represented the first of a new breed of aircraft. If things worked out right, Willard dreamed, then one way or another, he would become the Henry Ford of the aircraft industry.

By November 1953, the CCW-5's experimental airworthiness certificate expired. When the safety agent, E.R. Smith, signed off on the renewal inspection, he talked to Walt and Willard about their intentions for the odd-looking airplane, and Willard relayed Heinz' plan.

Smith shook his head and counseled, "getting a standard certificate for a new airplane design is very different from introducing, say, a new model car. Federal law mandates an exacting process by which a new aircraft can be certified by the CAA for manufacture, and it's not for amateurs. And you can't mass produce these for sale until it's certified."

"Give me an idea what's involved," Willard asked.

"You have a nice prototype here in the Five-Seven-Charlie," Smith said, referring to the plane by its abbreviated CAA registration number. "Now that you have a prototype, you need to get a Type Certificate. The Type Certificate means the CAA is satisfied that Five-Seven-Charlie conforms to all the civil air regulations for safe operation of the ship under every conceivable flight condition."

"And once I have a Type Certificate, then what?"

Smith continued, "Then you can start producing them for sale. Each plane you produce has to conform to the Type Certificate Data Sheet, which will be the specification for this Channel Wing aircraft. As long as every plane produced is produced according to the Data Sheet, it gets a standard Certificate of Airworthiness. It's not considered experimental anymore."

Willard frowned. "And how long does that take, to get this Type Certificate?"

"That's hard to say," Smith answered. "If you have a team of engineers and a test pilot who have managed to get Type Certificates for other planes, and they have everything they need to conduct the tests at the ready, and they can work at it full-time, and they don't run into any problems and nothing goes wrong, then I'd estimate you could get one in, say, six-to-twelve months."[9]

Willard's face brightened.

Smith looked hard at him. *"That's if everything goes right,"* he said.

Willard and Walt were both lost in their respective imaginations.

Turning to Walt, Smith offered, "I'll send you a copy of the regs. Then you'll know what you need to do."[10]

"Sounds good," Walt accepted with a smile. A week later he received the packet in the mail, and what he read worried him.

Obtaining a Type Certificate would be no trifle. According to the Civil Air Regulations in effect in 1953, the process entailed, first, a battery of ground tests to ensure that all the components of the prototype (down to the nuts and bolts) were designed and constructed in conformity to the regulations. So, for example, an engineer would have to test the gas tank to ensure that a full tank would not spring a leak when the aircraft flew at its top speed. Similar operating tests would have to be made of the wheels, tires, hinges, cables, brakes, pumps, valves, filters, radiators, carburetors, drains, switches, batteries, lights, instruments, belts, and radios under a variety of extreme conditions.

After successful ground tests, CAA test pilots would then conduct flight tests of the prototype during which they'd push the aircraft to its performance limits. For example, a pilot would test the CCW-5's conformity to performance regulations regarding normal operations such as takeoff, climbing, and landing both with and without engines, both light and overloaded. A pilot would also test how easily it recovered from hazardous situations such as spins and stalls.

At the same time, the CAA would require the Custer Channel Wing Corporation to provide detailed drawings of each part of the airplane, inside and out, with exact measurements and specifications, as well as an Airplane Flight Manual for the pilot, and a detailed maintenance program. In all, the effort to gain a Type Certificate would involve an exhaustive and expensive battery of tests, documentation, analysis and review—if all went well. If any issues cropped up with any of the items listed above, they would have to be resolved and retested, and would thereby lengthen the process. And each year, the CAA was adding to the regulations. Obtaining this certification for an aircraft of such a radically unique design would be that much more difficult. Some of the rules might have to be modified; some of the design specifications might have to be changed.

With the Type Certification test requirements in mind, Walt continued to work with the new prototype over the next year. Although "the experimental engines which were installed in the ship fell far short of developing the rated power for which the aircraft was designed," Willard wrote in his corporation's annual report, the engines they'd managed to procure provided enough power to conduct a long list of experiments. These helped Walt get a feel for the new craft's behavior in the air as well as in takeoffs and landings under different circumstances and in a variety of conditions.

While Walt continued experimenting with these less powerful engines, Willard envisioned how he could use a much more powerful jet engine.

Jet engines had been introduced to aircraft originally by Germany about the time of the Second World War, with a British model close behind. The rotary motion of the jet engine felt smoother than the pounding reciprocating pistons. The British shared their engine with the Americans, and its development accelerated after the war.

Like propeller engines, jet engines are designed to propel the aircraft forward. Unlike propeller engines, however, jet engines are designed not to turn a crankshaft, but to compress air. This compressed air is then heated, and the resulting exhaust exits the engine explosively, propelling the aircraft forward at speeds unattainable by propeller engines. The speed of the aircraft, as always, is the object.

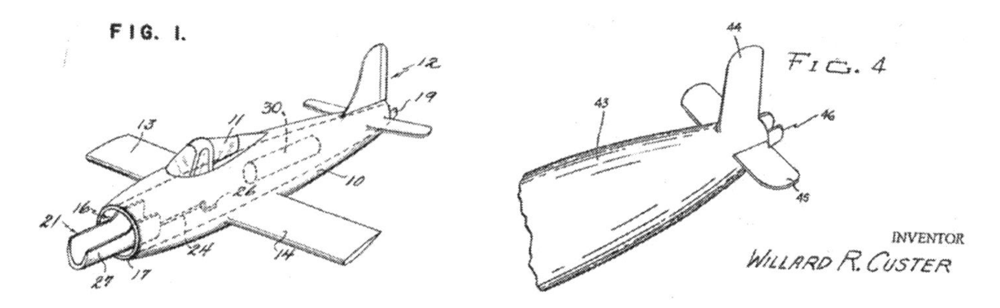

Figure 5.5 — Channel applications with jet engines: intake (left) and exhaust (right)

Willard, however, focused on the speed of the air entering and—especially—exiting the jet engine. He often used compressed air to demonstrate how fast-moving air, when shot across a screwdriver, for example, could make that object rise. The idea of the jet engine as an air compressor intrigued him.

By 1952, Willard had already patented Channel Wing designs that made use of jet engines, but these patents simply replaced propeller engines with jet engines, or they embedded the engines in the channel-shaped wing to bring the engine's intake and exhaust airflows closer to the wing surface. Beginning in 1954, however, he removed the channel from the wing altogether, and integrated the channel directly into the jet engine's intake and exhaust manifolds (Figure 5.5).[11]

With either of these designs, the channel would provide considerable static lift upon engine ignition. Willard believed it would enable jet aircraft to take off at 20 or 25 mph, eliminating their need for long runways. The channel at the intake could be extended or withdrawn, thereby employing it at takeoff and landing, but stowing it for conventional jet flight. The channel at the exhaust could be pivoted, changing the direction of the discharge and thereby the direction of the aircraft.

True to custom, Willard not only diagrammed his ideas for the patent, he also built models and working mockups. Later, in 1958, he would write in his corporation's annual report, "Great progress was made in research for the adaption of the channel wing to the modern jet engine, and we have now proved that the channel wing principle adapted to jet aircraft will give phenomenal results and will...permit aircraft equipped with channels to take off from and land in a small space, or to rise or descend more quickly, without sacrificing forward speeds."

Meanwhile, as Walt learned the CCW-5 with its less powerful combustion engines, he and Willard organized exhibitions for the press and other interested bystanders, showing off the CCW-5 in approximately one dozen public demonstrations.[12] They could be entertaining and awe-inspiring to watch, but they were costly and inherently dangerous. The CAA classified the CCW-5 an experimental plane. Though the agency ruled it airworthy, it presented a heightened risk, and the CAA tightly restricted where it could fly (over unpopulated areas), how far (typically, within a ten-mile radius), and who could be onboard (only essential personnel).[13] Lula later went one step further and demanded that only one of her sons—never both—could fly in the plane at the same time. She didn't want to lose both should there be a mishap. On top of all this, the maneuvers that Walt performed in these demonstrations were risky, leaving no room for error.

Because so much depended on this new prototype, and the previous NACA test report had been so disappointing, Willard and Walt carefully compiled their own corpus of test results, using the sky as their laboratory, and inviting public notice.

By the start of 1954, Willard and Walt reevaluated the effort necessary to obtain the CAA Type Certificate. However, they noted once again that the easiest solution would be a military contract. A military contract would relieve them of the need to obtain a Type Certificate at all because the military services were not subject to the CAA. All the demonstrating, prototyping, and testing would then be the government's responsibility. Short of that solution, however, their problem was that neither of them had experience in aircraft manufacturing or Type Certificate testing, and no one else associated with the effort did, either. Willard wrote to Frank Kelley at the time, "You know this thing is just about to break loose like a flood and it will take all those as close as you are to help steer a business as large as this is destined to be."[14] Frank, the only pilot on the board of directors and an excellent businessman, ran a photography studio. He did not have the background to manage the certification effort. Likewise, among Willard's Pittsburgh venture capitalists, none had expertise in aviation certification.

Looking outside the Corporation, Willard would have had a head start if Jack Baumann's Brigadier, the foundation fuselage of the CCW-5, had been certified, but it hadn't. At least, Willard had acquired Baumann's design drawings. His best help might have come from C.G. Taylor. He had obtained Type Certificates on his Chummy and Cub aircraft, but his own company's issues distracted him now, and Willard couldn't hire him, anyway. A nickel dropped into either company's piggy bank would ring as hollow. The Custer Channel Wing Corporation owed Willard $700 (seven weeks' worth) in back salary so it could squeeze out enough to pay other employees.

This is not to say that Willard's wallet hurt. He had decided early to keep his personal needs separate from those of his corporation. He had formed the corporation in 1939, during his surplus years, personally spending $200,000 on the venture, he estimated, while committing to a salary of no more than $100 a week, "because it's a struggling corporation, and I had to let it exist."[15] But in those early years, he and his older brother Cecil had also made investments together in real estate in Baltimore, Maryland, and Washington, D.C. Now, in 1954, he benefitted from some return from those investments, and his expenses were lower than they had been in years because his children were all grown.

But his corporation was another matter.

In May 1954, the Corporation had a total of $764 cash in the bank. In June, Heinz loaned the Corporation $1,000 and that year the Corporation recorded receiving a total of $15,000 in advances from him. Nevertheless, for resources to obtain the CCW-5's Type Certificate, as well as all the other operating costs incurred in testing, demonstrating, maintaining, and otherwise advertising the CCW-5, Willard and his corporation were forced to find additional sources of cash.

As if pawning his valuables, Willard sold the CCW-2 for $5,000, but he had no other product to sell to generate income. [16] Some corporate stock remained, Class B (no voting power), priced at $2.00 per share. The corporation still held eighty-thousand shares of it. A business would normally sell stock to provide a discrete amount of funds with which to build assets of cash-generating potential, such as a manufacturing plant, tooling and machinery. Willard, however, needed money to cover operational, maintenance, and testing costs until he held the Type Certificate in hand. So, the Corporation chose to sell some of its remaining stock.

In a corporate stock circular dated August 2, 1954, Willard represented his situation as follows:

> ...a prototype...known as the CCW-5 is currently being subjected to tests for certification at the Oxnard Airfield, California. An Approved Type Certificate issued by the Civil Aeronautics Administration is a prerequisite to the sale of an aircraft for civilian use. The issuance of such certificate is not assured but an application has been made and the requisite certification tests are ten per cent completed. The time required to complete these tests cannot be accurately forecast but it may reasonably be estimated at six to twelve months.

In the same circular, Willard estimated that the costs to complete the certification tests within that year would "not exceed $50,000." Whether he naïvely believed the estimates or chose to sugarcoat them for shareholders, hindsight proved him fantastically optimistic on both cost and time required. In either case, he knew that if he could land a military contract, the miscalculation wouldn't matter.

A Plane that Hovers

On Tuesday, August 24, 1954, the characteristically crystalline southern California skies formed a vault over the Ventura County Airport. Willard had invited a crowd to watch Walt demonstrate some of the CCW-5's unique performance features. Among the crowd were the CAA officials on his Type Certification board, Department of Defense representatives, a few newsmen, including William Austin (editor and publisher of *The SWATH*), and John Caywood, the airport director. Willard, who had flown out from Hagerstown with his son Reed, on leave from the Air Force, greeted each person individually and warmly.

Receiving clearance from the tower, Walt brought Five-Seven-Charlie (the CCW-5) onto runway Two-Five. Some in the audience were seeing the odd-looking aircraft for the first time.

Willard loved to watch people as they took in his plane, sometimes wide-eyed and open-mouthed, or smiling in wonder. He imagined the thoughts going through their minds. Adults tended to be skeptical, considering it (and him) to be an oddity, a mutant. But children and the more accepting were enthused, as if finding a new toy, eager to learn what a different airplane could do that an ordinary one could not. With both audiences, Willard was eager to show and tell, and he loved to see their minds open as he demonstrated his designs to them.

Five-Seven-Charlie pointed west, facing into an eleven mile-per-hour breeze slipping off the ocean. Walt opened the throttle and the engines sprang to attention, their roar echoing within the enveloping channels, the plane accelerating down the runway.

As the plane reached the 200 feet mark, Walt quickly pulled back with force on the yoke, careful not to bang the tail skid on the ground. The nose popped up, putting the aircraft instantly into a steep climb. Pulling the yoke still further, Walt turned hard to the left, spiraling upward in widening arcs, climbing as if drawn up the outside of a tornado funnel. Any other plane would have stalled instantly, but the CCW-5 kept climbing, rousing a grin from the pilot. Watching his altimeter, he leveled his 3,000-foot-per-min-

Figure 5.6 — Five-Seven-Charlie "hovers"

ute ascent when he reached 2,500 feet. He eased off the throttle to give the engines a short rest and to allow himself a moment to admire the gorgeous ocean view.

Reading his airspeed at 180 miles per hour, he slowed and began his descent. From a wide circle he positioned the plane into its base turn and lined up with the runway as if intending to land. One hundred sixty miles per hour, the airspeed indicator registered… then 120…then 90…then 60. His altitude dropped sharply approaching the runway, but Walt had no intention of landing. Instead, he leveled off. As the CCW-5 floated above midfield, Walt brought the nose up to a steep 22 degrees, slowing the craft still further while, counter-intuitively, increasing throttle, now sucking rivers of air through the channels (Figure 5.6).[17]

Geoff Sanders, a 10-year-old who lived nearby, remembered, "It was gosh-awful loud!"[18]

Airspeed sank below 40 miles per hour.

At the top of the instrument panel Walt had mounted a helicopter airspeed indicator because the airspeed indicator for a conventional twin-engine aircraft could not measure speed under 40 miles per hour. He eyed the helicopter airspeed indicator and he could not believe what his eyes registered. 11 miles per hour…0 miles per hour…5 miles per hour… 0 miles per hour again...then back to 11 miles per hour, the light breeze intermittently resisting the plane.

"I was out watching when Davidson brought the aircraft from a high speed to a slow, almost snail-like speed over the runway," Caywood, the airport director, told reporters later.

"I happened to be on the Oxnard airport that day," *The SWATH* editor later wrote. "I thought it was standing still!"[19]

"So I looked over the side," said Walt, afterward, "and saw I was practically standing still in the air. When I saw our people waving their arms at me from the field I knew we had something really great."[20]

A few days later, the Corporation issued a press release, and Willard captured national headlines once again:

> *New York Herald Tribune*, "'Tornado' Plane Hovers at 11 MPH"
>
> *Los Angeles Examiner*, "Plane 'Sucked Up' At 3,000 Ft a Minute, Lands at 11 MPH"
>
> *Oxnard Press Courier*, "Channel Wing Plane Hovers At Airport"
>
> *The (Baltimore) Sun*, "Wind-Lifted Roof Recalled As Custer's Plane Hovers"
>
> *(Charleston, SC) News and Courier*, "Slow-Speed Airplane Is Developed"
>
> *Philadelphia Inquirer*, "Barrel-Shaped Wings Slow Plane to 11 MPH"
>
> *(Hagerstown) Morning Herald*, "Channel Wing Craft Sets Another Mark"

Officially recorded at 11 miles per hour, the feat nevertheless ranked as "the slowest speed ever flown by a fixed-wing airplane." In a curious turn of events, the pilot who had hoped to capture first place in an international air race with the CCW-5 ultimately discovered his notoriety when the aircraft clocked an eleven-miles-per-hour crawl through the air.

The accomplishment and the associated publicity made the stock sale easy. The corporation brought in $156,000. However, the real prize never came in. No Department of Defense contract materialized, despite the attendance of military representatives that day.

"I simply cannot understand how our military brains can keep muffing this phenomenon," Willard complained to the Baltimore *Sun*, 30 August 1954.[21]

A Final Plea

When he was not demonstrating the CCW-5's capabilities in the air, Walt Davidson was promoting its possibilities on the ground. He had been in sales back in Pittsburgh, and he designed a ten-page brochure to market the "Kingbird," a corporate shuttle aircraft based on the CCW-5 but enlarged for comfort. It sported adjustable seats of spring and foam rubber, covered in gabardine, including one three-person seat in the rear. That seat tipped forward to access the luggage compartment, which was also accessible from the outside, behind the wing. Entrance to the temperature-controlled, sound-insulated cabin would be as easy as getting into a sedan. Even the instrument panel would be stylish and blend harmoniously with the luxurious look and feel of the interior. Walt had described his dream plane.[22]

One year later, a note sent from Albert Davis, the Corporation's Secretary-Treasurer, to Edward Summers, a stockholder in Hagerstown, would provide a synopsis of activity in Oxnard: "The CCW-5 is performing beautiful. [sic] We are flying most every day..."

But to Walt, activity was not the same as making progress. "I can't understand it," he confided over the phone to Frank Kelley one December 1954 morning. He had been working hard, but the results to date were nothing like he had envisioned. He groped for some explanation. "Everything is upside down," he said. "We put on these shows, and they're well attended. But some come just for the air circus. Everyone oohs and aahs, but where's the brass? Why isn't the military calling Willard, demanding that he name a price—any price—just so they can get their hands on it? Instead, they're sitting on their pretty white thrones, ignoring the man, when they ought to be down here worshipping him!

"It's like Willard says, Frank. The high and mighty—they don't get it. It's the common people who get it. But the common people aren't running the country."

"Heinz gets it," Frank countered. "He's upper class."

"I don't know if he does," Walt countered. "I'd sooner put my money on C.G. He's common man."[23]

Walt told Frank that he'd submitted an article on the subject to the Experimental Aircraft Association (EAA), a recently chartered organization of pilots and airplane enthusiasts. He and Frank had both joined, so Frank eagerly read Walt's "A Thought on Aviation" when it appeared in EAA's *Experimenter* newsletter the following month.

The article critiques the aviation industry's obsession with speed, and its use of its "government subsidy (our tax money)" to "engineer the airplane into a missile," in Walt's words. He censures "most of all of the aeronautical influentials of this country" for their tunnel vision and exhorts "us common-folk [sic]" to "take stock of the situation."

In taking stock, Walt points out that mankind has been rebellious against God since the garden of Eden, "fighting laws of nature" and "making multitudes of mistakes," the most germane to his article being that of copying the bird's wing without understanding how to use it. "Since 300 BC until 1900 AD, the bird-like gliders had caused death and anguish in thousands of men possessing the urge to fly," yet the Wright Brothers and the leadership of the 1950s' aviation industry had adopted the same perilous approach, "the wrong theory...the wrong road!" he warned.

According to Walt, then, God had recently intervened in history, having "chosen a Mr. W.R. Custer, a blacksmith and auto mechanic by trade... to get us back on the right road of aviation progress." He had in mind, of course, the Channel Wing aircraft, which he says is based on "nature's secret—move the air to get flight." His article ends with a personal and desperate plea to "you experimenters, homebuilders, and racing plan enthusiasts" to "change to the Custer Channel Wing now!"

Walt's plea rang urgent because he could sense his faith failing. His doubts were rising with the new debts. Most of the cash garnered from the stock sale the previous year had been used to retire debt, with nothing saved to pay this year's bills. The $50,000 estimated to complete the Type Certificate had been completely spent for that purpose, but little progress had been made.

Then Willard received word from C.G. Taylor that Taylorcraft had folded. C.G. had spent the previous five years trying to resurrect his BC-12 side-by-side aircraft, but the company had done poorly. There is no indication that he ever started work on the Channel Wing "liaison" aircraft that he had intended to design and develop. Considering C.G.'s experience in aircraft production and Willard's lack thereof, a Taylor-Custer business partnership might have been perfect, but C.G. could not bring a viable company to the deal. So, he returned to Ohio to work for Product Designs, Inc., the family's business near Alliance, where he could concurrently work a variety of small projects while enjoying some financial security.[24] In Ohio, he would continue sketching on his idea for a side-by-side two-seat Channel Wing. It would be a hobby until an investor materialized.

What would be the final demonstration in California took place on December 8, 1955. It serves as an example of Willard's desire to properly document the demonstrable capabilities of his aircraft. To verify and add credibility to his growing body of test results, he invited a disinterested third party, Ray and Ray, an aviation operations and research firm in Washington, D.C., to observe the CCW-5 in flight and to document its observations. Two weeks later the firm published its short report. On a day reported to be 72 degrees Fahrenheit with clear skies, James Ray had observed the CCW-5 flying at 22 miles per hour. Noting that a conventional airplane would have fallen out of the sky at even 72 miles per hour, Ray calculated that the lift in the channels alone held nearly 3,900 pounds in the air:

> The aircraft cleared the ground in 221 feet against a five mile breeze; it climbed at an angle of about 12 degrees and attained a minimum level flight air speed of approximately 22 m.p.h., which is more than 50 m.p.h. slower than its calculated stalling speed...At the observed speed of 22 m.p.h., the calculated wing area would be lifting only 351 lbs. at its maximum lift coefficient. Thus the remaining 3,899 lbs [the plane weighed 4,250 lbs.] was being carried by the channels.[25]

Finding a Market

Nearly a year later, in September 1956, a blowout market for the Channel Wing erupted half a continent away in McAllen, Texas, at the very southern tip of the state. As if responding to Walt's article, Louis Leone announced to his local newspaper that he would be building Channel Wing aircraft. Leone developed real estate in McAllen, not airplanes, but he led a group of investors from the Texas and California cattle and oil industries. These men were interested in the Channel Wing's slow flight and short take-off and landing capabilities for use in remotely located cattle ranches and oil rigs. Because the commercial aircraft industry focused too much on flying faster with larger payloads to be interested in building Channel Wing aircraft, these customers would build their own. They had formed a corporation—the Custer Channel Wing Manufacturing Company—and had already purchased the required building, tools, and equipment from the former Regent Aircraft company. They were reportedly poised to "go into immediate commercial production" of two Channel Wing models: the CCW-5 and a two-seat model.[26]

According to *Jane's All the World's Aircraft: 1956-57*, Willard called the two-seat Channel Wing the CCW-2, forever to be confused with the CCW-2 previously tested by NACA.[27] Albert Davis, the Corporation's Secretary-Trea-

Figure 5.7 — 1948 Air Trails Pictorial *cover's concept single-engine CCW-2*

surer, in his report the previous year describing the CCW-5's activity, had also reported that the "program on the CCW-2 is coming along nicely and we hope to have the first one to test fly in about 60 days...We have so many orders it will be several years before we can possibly get caught up."

Willard had sketched and modeled this prototype years before (Figure 5.7).[28] The crop-dusting community saw this aircraft as the perfect single-engine duster, capable of drifting slowly over crops and landing in a cow pasture. Oil men, ranchers, and foresters wanted it to patrol long stretches at slow speeds and to land in remote areas. They could use helicopters, but helicopters with troubled engines like to plummet like rocks while troubled airplanes will still glide. Over time the aircraft could evolve into a flying car, taking off and landing in the family's backyard, folding its wing stubs on landing, and parking in the driveway.

Whether Leone appreciated the amount of "additional flight testing" that he said had to be done in McAllen prior to moving "full speed ahead," his group's announcement highlighted a growing Channel Wing demand. As news of the CCW-5's capabilities spread, interest in other applications rose, as well, even from areas outside the United States, including "Africa, South America, Canada—the underdeveloped frontiers of the world with rudimentary facilities and limited clear space." Island nations with similar limitations on landing strips, such as Japan and Caribbean nations, were also becoming interested and visiting the Corporation's President, Willard Custer.[29]

As Albert Davis' report reveals, Willard is already taking orders. He has tapped into an appetite for Channel Wing aircraft even though he is unable to immediately satisfy the growing demand. The CCW-5 demonstrations have sold his aircraft, even though he does not have the necessary Type Certificate to mass produce either a CCW-2 or a CCW-5. His is a calculated risk. In his mind, a stack of orders is good bait to lure the existing manufacturers, hungry for the opportunity to fulfill those orders. And it has worked. Leone's group has taken the bait, but they do not have the expertise themselves to bring in a Type Certificate. Who can do that, how much longer will it take, and how long will these customers wait before they lose confidence?

Finally, Leone claimed in the article to have licensed manufacturing rights from Willard's corporation, assuming the Type Certification testing successfully concludes in Texas. Finishing the certification testing would involve Willard's moving Five-Seven-Charlie and some part of his operation to McAllen. While a move to Texas probably did not appeal to Willard, he admitted to himself that operating out of California, on the opposite coast of the continent, certainly had its drawbacks. However, the notion of relocating his base of operations was not out of the question, if he found that a compelling opportunity might be had. So, Willard began to explore the possibilities.

Akron, Ohio

As he entered the six-story, red brick headquarters of Goodyear in the heart of Akron's downtown, Willard felt a sense of deja vu. He recalled entering the Briggs Manufacturing headquarters in Detroit seventeen years earlier, preparing to meet Howard Bonbright. This day, in early summer 1957, he would meet Pete D'Anna, head of Goodyear Aircraft Company's Advanced Programs. Willard again wore his three-piece black pinstripe suit, fedora, and best tie, but this time he didn't carry a film of a model to show. He had a bonafide Custer Channel Wing prototype sitting in a California hangar to discuss.[30]

Willard had been familiar with Goodyear's blimps, but he'd learned that Goodyear also had an aircraft company that had manufactured Corsairs during the war, as well as its own amphibious aircraft, the "Duck."[31] It was now experimenting with an inflatable rubber aircraft project (GA-33 Inflatoplane) for the Army. This might be the kind of innovative company, like Taylorcraft, that would be interested in leveraging its facilities and government relationships to develop a Channel Wing aircraft of its own design.

Willard had set up the appointment simply to be introduced and to explore whether they had anything in common as a basis for future discussions. They met for lunch, informally, in Goodyear Hall's employee cafeteria. Willard overdressed intentionally. He knew he'd be doing most of the talking, so he chose merely a slice of lemon meringue pie and coffee with cream and sugar, and picked up the tab for Pete, as well.

As they seated themselves at the remotest empty table, Pete apologized for the setting, but his days were otherwise very busy with meetings. Willard said he understood; he had to hurry on, too. They chitchatted about Akron and the splendid weather that year. Then Willard began his story.

Willard told Pete about the barn roof flying off in the thunderstorm, and the genesis of his concept that the crux of the matter in aircraft lift is not the airspeed of the craft, but the speed of the air coursing over the wing. He told him about the Bumblebee and the Wright Field tests in nearby Dayton. He showed him pictures of models to spark his imagination. Then he fast-forwarded to the CCW-5 and its recent achievements. He showed him his company brochure, and pictures of his two-seater already in demand. He also showed how channel wings could be put on any kind of aircraft.

"When can I see the meat of your test data?" Pete asked.

Willard hated the question. He found it both frustrating and insulting. He had conducted three kinds of tests, none of them being the kind Pete sought. Primarily, Willard had experimented with different materials, propellers, and designs to find the best ways to achieve the aircraft performance results he wanted, and he had yet to be satisfied. These results he would never share because they constituted company secrets. Secondly, he had demonstrated the aircraft's STOL capabilities in public exhibitions, and these results were captured by the journalists or officials who attended those demonstrations. Willard considered those results public knowledge. Did Pete not read the front page of the newspaper? The third kind of tests were those being conducted to gain the Type Certificate. They were mundane, tedious things, and they demonstrated that the aircraft would fly as well as any conventional airplane without cracking up, not to demonstrate its STOL capabilities. Willard knew Pete craved performance test results like those formerly produced by Wright Field and NACA. Those results were old and applied to experimental models, now retired. Willard had no recent scientific studies of the CCW-5, with neatly tabularized results of trials using dependent and independent variables represented by Greek letters on pretty graphs, summarized and leading to favorable conclusions. He did not have the money to provide the staff nor the facilities to conduct such a study—nor did Wright Field have any interest in conducting one (he had inquired)—and Willard had so far found no way to solve that problem.

Willard did offer the report documented by Ray and Ray. Pete didn't care to see it.

As Willard described Five-Seven-Charlie's feats, he impressed Pete with his grasp of aerodynamic principles. Pete never would have guessed that Willard had not completed high school. But Willard could not supply what Pete wanted in documented test results.

"Before I can recommend that the company commit to this as a viable program," he insisted, "I ought to know a little more about what your airplane can do." In Willard's mind, however, the two things must happen in the opposite order: first, Pete should commit to the program, and then, when he had skin in the game, he'd be funding his own tests.

When he asked if Willard could conduct some tests for him in Akron, Willard repeated that he had no money. Pete suggested ways he could cut the costs of the tests, but the idea ruined Willard's lemon meringue pie. My gosh, Willard thought, he held the key to bringing every man, woman and child into the Air Age with the only safe, slow-flying aircraft in the world. He would not be put in a position where he had to beg anyone to believe in his demonstrated claims. He'd had enough of that with the US Patent Office.

In the end, they parted on friendly terms, and would indeed have a second meeting. But Pete found Willard less than forthcoming, even if very pleasant to talk to.

Willard, too, left disappointed in the conversation. He pegged Pete as a very good program manager, but Willard didn't need a manager. He craved another C.G. Taylor who could take the Channel Wing idea and make it his own. Besides, a rubber Channel Wing didn't much appeal to him.

Nevertheless, Pete generously offered Goodyear's facilities and access rights at the nearby Akron Fulton airport for Willard to use in CCW-5 Type Certification testing. Goodyear's blimp construction hangar and air dock anchored that airport. Small, with one runway, it sat low, as if in a bowl, giving it a semblance of privacy and permitting very little air traffic.[32]

That got Willard thinking. Every time he thought he might offload the Type Certification effort to another company, it came back to him, like a stray cat he had mistakenly fed at one time. The manufacturing company in McAllen, Texas sat poised and ready to help, but it offered little towards Type Certification. If Willard must continue with the task, the timing and circumstances in Akron might be more favorable than in McAllen. Typically, the CAA gives an applicant three years to complete the process. Three years had passed since Walt and Willard had started the Type Certification, with limited progress, so they already had to file for an extension or a restart.

Likewise, the regional CAA officials oversee the process and evaluate the test results, so every relocation of the aircraft necessitates a start-over with a new CAA team. In Akron, Willard would meet with a new team of CAA officials to restart the process. Akron would be cheaper, and it was closer to home than California or Texas. Furthermore, Akron sat at a hub of aviation activity, including the U.S. Naval Air Station, and it sat close to C.G. Taylor and to Wright Air Force Base, besides. With all of these advantages, Willard decided to move his CCW-5 operation to Akron. He would have a representative there while the corporation remained in Hagerstown.

Frank Kelley cheered the decision because Akron sat next door to his hometown in Canton.

Walt Davidson, on the other hand, took it as a sign from heaven that he should leave the Channel Wing endeavor. He would not uproot his family again, especially when he could not depend on a salary. He broke off his relationship with the Channel Wing camp instantly and completely.[33] Curley, Willard's son, began piloting Five-Seven-Charlie as they prepared for the move. While Curley had a full-time job and a family of his own in Hagerstown, he could take some time off from his job and donate some time to his dad, even for no pay.

Without delay, Five-Seven-Charlie left its California hangar to undertake what the company's 1958 annual report described as "a cross-country flight to twenty cities scattered from California through the Southwest and into Mexico and up to Ohio." It turned into a tour packed with CCW-5 demonstrations.[34]

In film footage from a Shreveport, Louisiana, demonstration, a twin-engine Beech aircraft takes off ahead of the CCW-5. In contrast to the Beech's long takeoff run, the short hop of the CCW-5 is striking. Five-Seven-Charlie appears to leap into the air as if escaping bindings. Walt Boyne, the aircraft historian and writer, saw the films, and wrote:

> These demonstrations consisted largely of maximum performance take-offs, with a terrifyingly steep climb out, using steep turns to the right and left at a high angle of attack and at speeds well below the stall speed of conventional aircraft to impress the audience. Motion pictures of the demonstration are heart-stopping; you know that an aircraft in that attitude, at that altitude and airspeed, is going to crash—but the CCW just keeps on turning.

From Shreveport, the CCW-5 flew to Reynosa, Mexico, just across the

Rio Grande from McAllen, Texas, for a demonstration before "high Mexican Government officials."

Next, the plane flew to Enid, Oklahoma, home of two airports including Vance Air Force Base, for more demonstrations. Willard spent three weeks there in August 1957, speaking to the Enid Lions Clubs, local chapters of the Experimental Aircraft Association, and other civic organizations, and to impromptu assemblies, as well. Those who attended described him as exciting and entertaining. He had become a master at demonstrations, large and small. For example, when he appeared on the TV game show "I've Got a Secret" (a clip can be viewed on YouTube),[35] he demonstrated the principle of the "speed of air" in the Channel Wing design by using a simple sheet of writing paper. Watching it now, one realizes Willard had done this demonstration a thousand times, but it is so simple, and it communicates so well that it is exquisite. Another example: Willard would put a screwdriver, blade pointing down, in his shirt pocket. He would point out that a screwdriver has no aerodynamic properties whatsoever. But then, from a pressurized air hose, he would release a stream of fast-moving air directly above the screwdriver handle, and the tool would rise from his pocket and, as if by magic, dance in the air until Willard grabbed it with his free hand.

He capped his stay in Oklahoma with a demonstration at Woodring airport before an audience estimated at 200 people. The former mayor and present city councilman, Sam Stoner, became so interested that he eventually joined the Custer Corporation's board of directors.

The tour ended at the Akron-Canton Airport. There, as recorded in the FAA records, Willard obtained hangar space, and began to evaluate some "new style propeller blades." The busy Akron-Canton Airport sat high on ground south of the city. Willard also accepted Pete D'Anna's invitation and made use of the flight test area dominated by Goodyear at Akron Fulton Airport. There he restarted flight evaluation tests required to obtain CAA type certification.

At 58 years of age, and with the CCW-5 hangared in a new home, Willard needed a turn for the better. The Custer Channel Wing Corporation could not afford to pay him, and he and Lula were eating only what they could grow in their garden or gather from their chicken coop or shoot in the field. The children were all grown and married now, with families of their own. But if Lula ever complained, the children never heard it.

I do the very best I can, I mean to keep going. If the end brings me out all right, then what is said against me won't matter. If I'm wrong, ten angels swearing I was right won't make a difference.

– Abraham Lincoln

VI The Stockholders 1958 - 1963

This case revolves around the effort of the officers and directors of a small corporation finally to bring to fruition the ultimate purpose of the corporation which had been in the making for more than 20 years. While the effort involved the possibility, even likelihood, of personal loss to themselves, it offered them no possibility of personal gain other than that to be shared equally by all stockholders in the event the effort was successful.[1]

- Seegmiller, Wilner, & Custer

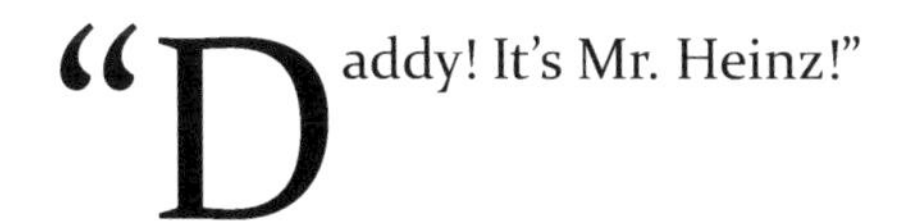

“Daddy! It’s Mr. Heinz!”

Helen had left Willard’s kitchen to answer the phone on the little table in the foyer. She and Willard were working in the warmest room in the house that January weekday in 1958, gathering 1957 financial records for the accountants.

Willard took the heavy black receiver from her and covered the mouthpiece. Pausing, as usual, long enough for her to climb the stairs and quietly lift the receiver in the bedroom, Willard put the phone to his lips.

“Happy New Year, Mr. Heinz,” he said, enthusiastically.

“Happy New Year, Willard. How was your holiday?” Jack Heinz was grinning, Willard could tell.

"Very nice, sir. Hagerstown has avoided the snow so far this year. I bet you can't say as much there in Pittsburgh, eh?" Willard chuckled.

"No," Heinz chuckled in return, "that's why I don't winter in Pittsburgh. I'm actually in New York at the moment. I've been talking to some people about your situation, Willard. We have some ideas I want to run by you. I think you'll be excited about them. Do you have a few minutes?"

"Of course, Mr. Heinz," Willard answered, wondering. "Let me get a pencil and paper."

"Of course."

"And do you mind if I have Helen listen in?"[2]

Heinz described his idea to form a second corporation that would be a subsidiary to the Custer Channel Wing Corporation. The new corporation would focus on licensing, sales and marketing, which is to say, turning a profit. Willard, however, would focus on the Hagerstown corporation, engineering activities, research into jet engine applications of channel wing configurations, testing, aircraft performance, and domestic and international patents. This also included wrapping up the FAA Type Certificate. (The CAA turned into the FAA on August 23, 1958.)

In addition to the separation of concerns, the arrangement would also include a corporate restructuring involving the transfer of money, reassigning debt from Hagerstown to the new company in exchange for certain patent rights, changes to the Hagerstown corporate charter, a forty-for-one stock split requiring a certificate exchange, and more. Heinz would also retitle Five-Seven-Charlie, making the subsidiary the new owner of that aircraft. Willard cared most that he, Willard, would maintain controlling stock ownership in both companies, thus maintaining control of his patents. Heinz cared most that his investment in Five-Seven-Charlie may yet pay a dividend.[3]

Despite any misgivings lingering from his big corporation experience in Detroit with Briggs Manufacturing, Willard agreed to the idea as the only real opportunity open to him.

By April 7, 1958, the Custer-Frazer Corporation had been incorporated in Delaware, a subsidiary to the Custer Channel Wing Corporation (in Hagerstown), with its headquarters in New York City. As Chairman of the Board, Willard led a Board of Directors comprised of Pittsburgh and New York members, including Jack Heinz. Barr Peat, who had introduced Willard and

Heinz, held one of two vice president seats. Willard had not yet met the president of the company, Joseph W. Frazer.[4]

Frazer was a retired automobile company executive. He'd led Chrysler as General Sales Manager in the 1920s and 1930s when Chrysler sold a million cars per year. He'd gone on to preside over two more automobile companies in succession, leading them to prosperity, according to his resume. Seeing an opportunity with the Channel Wing aircraft, he came out of retirement to lead the Custer-Frazer group. How well his automotive background would translate into experimental aircraft remained to be seen.

For his part, Frazer quickly produced impressive results, as published in the Custer Channel Wing Corporation's 1958 Annual Report:

> ...on October 2, 1958, an agreement between the Custer Frazer Corporation and William G. Spence of Montreal, Canada, was signed, under the terms of which the facilities of Noorduyn Norseman Aircraft, Ltd., of Canada, in combination with the Leader Products, Ltd., and the Gorde Tool and Die Company have been pooled to carry out the manufacture of the CCW-5 on a large scale basis. The agreement calls for the creation of a Canadian corporation to be known as the Custer Channel Wing (Canada), Ltd.

Frazer anticipated production would begin shortly in Montreal, in the spring thaw of 1959, for CCW-5s being offered at $55,000 each. Involving an international company pleased Willard, who hoped to make American aircraft companies jealous and sorry that they had not adopted his aircraft design.

In addition to CCW-5s, the Canadian business modeled at least one of its own designs: a ten-seat commuter aircraft based on the CCW-5 and called the CCW-10 (Figure 6.1).[5]

The Custer-Frazer group would leverage the production capacity of the Canadian plants to satisfy U.S. orders, and the Canadian company could also sell aircraft in Canada. Custer-Frazer committed to an initial production run of one hundred CCW-5s for its U.S. market and would receive a cut on all Canadian sales.

The Hagerstown corporation's 1958 annual report focused entirely on the corporate re-organization. "Accordingly, much of this report is devoted to the new organization and facilities recently established for the purpose of successful marketing, by one means or another, the inventions embodied in the Custer patents," Willard wrote.

Figure 6.1 — Canadian CCW-10 model

The report further claimed, "Manufacture of ... aircraft is to commence immediately, with delivery of the first aircraft to be in mid-1959... For the first time in the history of your Company, we are in a position to accept firm orders for the delivery of Custer Channel Wing aircraft."[6] This statement rings partially true; the Corporation could accept orders, but it could not deliver on them. In principle, this arrangement resembled the McAllen venture. Without FAA certified data sheets, the company had no machine specifications by which to produce Channel Wing parts, and the machines could only sit idle.

But "idle" would never describe Willard. A few months later, in August 1959, he updated the Corporation's stockholders on "our recent and current activities."

First, he wrote, the Canadian company was "progressing very well... Financing...has been substantially completed...The necessary engineering for the manufacture of the CCW-5 as a production aircraft has been practically completed...and our Canadian licensee will be able to deliver aircraft in June of 1960..." In other words, production has been delayed a year, but the company is almost there.

Second, and most pertinent to the success of that effort, a new test pilot had been identified for "thoroughly testing our prototype as a part of the certification program for production models." Willard described 39-year-old Dick Ulm, still employed as the test pilot for Goodyear's rubber Inflatoplane, as the new test pilot and a "flight engineer for the Federal Aviation Agency. Ground and flight tests are now being conducted with very gratifying results." Even better, Five-Seven-Charlie had, at long last, gotten its upgraded engines. Seven years earlier, Walt Davidson had grumbled that the Continental engines were not powerful enough. Now, at long last, the upgraded engines had arrived and been installed. There could be no better news. But Willard had more.

In the report, Willard described Mr. Frazer as "most active" and "an aggressive minded tireless worker dedicated to our success." His domestic and international licensing negotiations were "progressing most favorably."

Finally, in a lobbying blitz, the Corporation had presented "our program and the Custer Channel Wing principle" to members of the Armed Forces Service Committee in Congress, "members of the Senate," the FAA, and officers of the Air Force, Army, Navy, and Marines. For the first time, the "program has been presented in a detailed and authoritative manner to substantially all potentially interested segments of the federal government." The next step was to demonstrate Five-Seven-Charlie to a collective audience of interested departmental representatives.

That opportunity materialized quickly. The Pentagon's top brass scheduled the demonstration for September, at the airfield at the Quantico Marine Base in Virginia, just south of Washington, D.C. The airfield there was the home of the Marine Corps' helicopter program. Willard saw it as his chance of a lifetime. But for whatever reason, Dick Ulm could not fly. Willard scrambled to find a top-notch demonstration pilot, or he risked a nightmare come true. This time he got lucky.

Quantico, Virginia Demonstration

Willard's new connections in Montreal now paid off. Canadian Bill Atrill had been an instructor pilot in both the Royal Canadian Air Force and the Montreal Flying Club. After his military service, he'd involved himself in business aviation. Well reputed in Canadian aviation circles, he had piloted planes of all sizes, from the Piper Cub to the DC-3s, and his logbooks show he'd flown more than fifty different types of aircraft. Bill Spence, President of Custer Channel Wing, Canada, played matchmaker between Willard and his new pilot.

Bill Atrill met Willard on July 27, 1959 at the hangar in Akron. That day he flew Willard to Hagerstown and back, but not in Five-Seven-Charlie. The flight served as Atrill's job interview and his introduction to the company. Willard hired him, and Atrill returned to Akron two weeks later, on August 12, to begin learning the intricacies of flying the CCW-5. He would have to learn fast.

In the Corporation's Annual Report for 1959, dated August 16, Willard wrote, "In the next few weeks, the company plans to make a comprehensive demonstration of [the] CCW-5 prototype before all of the armed forces, committee members of Congress and the Senate, air attaches and NATO embassies, representatives of aircraft manufacturers, both foreign and domestic, representatives of the Federal Aviation Agency and the Aviation Industry Association, and members of the press and wire services. The results should be, after successful demonstrations, recommendation and full acceptance of the Custer Channel Wing."

Figure 6.2 — Pilot Bill Atrill

Clearly, Willard expected the upcoming demonstration to persuade any skeptical minds. And, no one more than the newcomer, Atrill, felt the pressure to turn in an impressive performance. Which is why the timing was especially merciless when, on August 22, as he was preparing to land Five-Seven-Charlie, Atrill suspected the plane's landing gear was not working.

He radioed the airport's air traffic controller.

"Tower, this is Five-Seven-Charlie. Request clearance for a fly-by to get a visual on my gear. I don't feel my wheels. Over."

"Five-Seven-Charlie, this is the Tower. You are cleared for a fly-by. Maintain this heading and descend to 100 feet for a visual. Over."

"This is Five-Seven-Charlie. Roger that. Over."

Atrill had just passed the tower when his radio came to life again.

"Tower to Five-Seven-Charlie. You are gear-up. Repeat. You are gear-up. Do you copy? Over."

"Five-Seven-Charlie. Copy that. I will try to shake it loose on this circuit. Failing that, will request clearance for a gear-up landing. Over."

Regaining altitude as the air traffic controller cleared the skies and runways of traffic, Atrill tried several maneuvers that use gravity or centrifugal force to yank the gear down. None of them worked. So, he was left with no choice; he had no wheels. He would have to land the plane flat on her belly.

A seasoned pilot, Atrill was annoyed, but not worried. Had he been landing a conventional twin engine aircraft, the prospects for a harmless outcome would have been more remote. In that scenario, the propellers would strike the ground first, destroying them. The engines might directly impact the ground with the fuselage, and the wings might suffer structural harm. Coming aground at fifty or sixty miles per hour would cause a long slide with decreasing ability to steer. The damage would almost certainly render impossible any demonstration scheduled three weeks later.

But Atrill had already learned enough about a Channel Wing to know that it had several distinct advantages in this kind of emergency. So, he brought the plane down slowly and reduced speed to 30 miles per hour (half the speed of a conventional aircraft). He lined up with the grassy Runway Safety Area that bordered the asphalt runway. Then he set the channels and fuselage down on the turf gently, like laying a baby in a crib, and cut the engines. The plane slid across the smooth ground to an easy stop. The propellers and engines never felt a thing, safely encased within the channels.

Figure 6.3 — Pilot Bill Atrill leans on the wing after a gear-up landing

Atrill emerged from the cockpit as four mechanics ran over to inspect any damage. The fire hose lay unneeded. Airport personnel jacked up the plane, manually lowered the landing gear, and then towed her to the hangar. Three days later, her landing gear mechanism repaired, her scratches and grass stains gone, and the paint restored for her upcoming demonstration, Five-Seven-Charlie was flying again, as if nothing had happened.

Through the remaining few days of August and the first eleven days of September, Atrill practiced with the plane almost daily. The demonstration at Quantico would be on Tuesday, September 18th. On the 12th, the preceding Wednesday, Atrill and a co-pilot flew the CCW-5 from Akron to Hagerstown. They refueled and proceeded to the small Eastern West Virginia Regional Airport at Shepherd Field just south of Martinsburg, West Virginia, near where Willard had grown up. From there they flew the plane to Quantico, Virginia.

With the few practice days that he had in Quantico, Atrill practiced some "aerobatics" in other aircraft. Then on Monday, the day before the big demonstration, he and Willard settled on a demonstration script, and Atrill, as if cramming for a final exam, took his last few practice flights in Five-Seven-Charlie to apply what "aerobatics" he had just learned.

The airfield at Quantico Marine Base squeezes into a slim triangle of shoreline formed by the Potomac River on the east, the Chopawamsic Creek on the south, and the Washington-to-Fredericksburg Railroad on the west. A single runway (02/20) points just 20 degrees east of true north, paralleling the track. Hugging the northern end, a cluster of five cube-shaped buildings of darkening red brick include two hangars, recognized by their wall-to-wall windows set in hunter green frames. A control tower and a headquarters building, of the same brick, complete the airfield.

Tuesday dawned sunny and clear. By late morning, the temperature was 65 degrees. Winds were from the north/northwest and variable at 12 MPH. Perfect flying weather.

At noon, the Deputy Secretary of Defense, Thomas S. Gates, Jr., and various dignitaries from the three military services, assembled on the temporary aluminum bleachers set up just inside the entrance gate. After being introduced, Willard, now 60 years old, stood before this august audience, leaning on his cane, and speaking over a loudspeaker. Poised on runway 02, the CCW-5 warmed up. Willard began by explaining to his audience the fundamentals of how his airplane worked, the crucial distinction between airspeed and the "speed of air" to obtain lift, and he previewed for them what they were about to witness.[7]

First, he announced, they would see the airplane lift off in fewer than five seconds and in less than 200 feet of runway. Next, they would see it climb at a steep 2,500-feet-per-minute incline. After that, the pilot would demonstrate the airplane's "hover" maneuver. And finally, the demonstration would end with an incredibly slow landing at fifteen MPH. Willard turned to Atrill and signaled him to prepare for takeoff. Then he silently prayed for a miracle.

The CCW-5 engines quickly increased to a roar. Atrill noticed the wind blowing perpendicular to the runway, but it was of little concern. In fact, it was perfect for hovering and slow flight.

Producing a stopwatch, Willard signaled Atrill to begin.

Atrill released the brakes, and a Navy reporter recorded what happened next.

Willard counted over the loudspeaker from a stop watch—'5-4-3-2-1'—by the time he intoned 'one' the 4,900 pound load lifted off before using 200 feet of runway. Atrill banked hard to the right as he climbed, looping out over the Potomac River.

The rest of the demonstration, which involved several takeoffs, progressed almost flawlessly as prescribed except that one of the takeoffs had to be aborted and redone. At its conclusion, Willard asked Atrill to step out of the cockpit and take his bow. Both were extremely pleased with his and Five-Seven-Charlie's performance.

Because of the successful demonstration, the Air Force's Research and Development Command (ARDC) made plans to further study the aircraft in the following months, giving Willard time to select a better brand of propellers. Then, according to Aviation Week, "the aircraft will be instrumented by the National Aeronautics and Space Administration [NASA, formerly NACA]. Flight testing probably will be conducted at a civil field at Canton [Akron], Ohio. Results of the ARDC investigation will be made available to all three military services."

Fine, but this was not the miracle Willard had hoped for.

A House Divided

Meanwhile, Frazer's licensing negotiations had borne fruit in the form of yet another Channel Wing corporation. Channelair, Inc., with offices in New York City, would be licensed "for the purpose of assembling and selling...the CCW-5 in the eastern states of the United States and islands of the

Caribbean Sea," according to its one report to stockholders. The parts to be assembled were to come from Custer Channel Wing of Canada. Curiously, Joe Frazer, while President of Custer-Frazer, would also chair the Channelair Board of Directors.

Unlike Frazer, Willard held no interest in Channelair, and neither did any other Channel Wing corporation. So, Frazer's stake seemed questionable to Willard, and he needed a lawyer to review the language in the proposed licensing agreement.

Cecil E. Custer, Willard's brother, had been admitted to the Washington, D.C. Bar in 1937. Prior to that time, he had been a finance and budget officer for the Civil Service Commission for five years, and he continued with that government agency until he retired in 1952 and went into private law practice. He had also been town counsel for the town of Gaithersburg, Maryland from 1956 to 1959 where he helped devise the town's zoning plan. That he had accounting skills and regulatory experience, then, is undeniable. However, Barr Peat, and probably others, questioned whether Cecil had sufficient corporate or contract expertise.

In 1959, Cecil became partner in Seegmiller, Wilmer, and Custer, a law firm in Frederick, Maryland. His clients included his brother's Channel Wing corporation. So, Willard asked Cecil to review the Channelair documents, and Cecil, accordingly, investigated whether the arrangement with Channelair promoted his brother's best interest and that of his Hagerstown corporation.

Cecil learned that Channelair had initially agreed to assemble parts manufactured in Montreal by Custer Channel Wing (Canada). Of course, there were no parts coming out of Canada because there was no Type Certificate Data Sheet by which to document any specifications. But all Channelair knew was that the Canadian supplier was failing. So, by June Channelair had discovered another company's "sandwich construction" process that would, it thought, be much cheaper. By August, Channelair had also designed its own single-engine, four-seat Channel Wing aircraft, and by October began seeking a license from Custer-Frazer to build it using its "sandwich construction" parts. Channelair wanted exclusive rights to that license for five years. In that five years, Custer-Frazer would reap license fees, but Willard and his Hagerstown company would see no other profit from the arrangement.

According to Cecil, the agreement "would have the effect of transferring to...Channelair, Inc...all production rights under the channel wing inventions and letters patent on propeller-driven aircraft without any [up front] payment to Custer-Frazer Corporation therefor."

So, Willard informed Frazer that he objected to the exclusionary language. The ghost of Howard Bonbright had returned to haunt him. He had invested his entire life in this aircraft; he refused to let Frazer reap profits from it without sharing them with him and his Hagerstown shareholders. Custer Channel Wing Corporation in Hagerstown held eighty-five percent of Custer-Frazer stock, and as the President and majority stockholder of that corporation, Willard rejected the agreement. He also began to doubt Frazer's loyalty.[8]

But Frazer would leave little room for doubt. He had never joined this game out of devotion to Willard or love of his aircraft; he just wanted to make money. Convinced that Willard had mismanaged opportunities with the McAllen group in Texas (before Frazer had even become involved) as well as the Montreal company, Frazer had purposely developed his own opportunity outside of Willard's direct control. Now, he and the entire Custer-Frazer board were determined that Willard would not mismanage yet another potential money maker. So, on Thursday evening, November 19, they voted behind Willard's back to ratify the Channelair license agreement.

How Willard found out about it is unknown, but the very next day, enraged at the double-cross, he sued Custer-Frazer in U.S. District Court for the Southern District of New York, as well as the U.S. District Court for the Western District of Pennsylvania at Pittsburgh, for "corporate waste and self-dealing by directors." At last the pent-up bile spilled, and in return, the defendants accused Willard of "capricious and obstructionist activities which...have for years prevented the commercial exploitation of the invention, to the prejudice and detriment of all the individuals and corporations involved."[9] Lost in all the finger-pointing and frustration loomed the real challenge to these two companies, the failure to gain the FAA certification. Heinz, Fraser, and the whole cadre of Pittsburgh investors could rightly accuse Willard of sabotaging this particular agreement. But, without a Type Certificate, this agreement represented a firm hold on an empty money sack.

In court, Willard opposed the entire Custer-Frazer board. He added his Hagerstown corporation as co-plaintiff, bringing in Helen and her voting shares to further reinforce his controlling shares. He now wielded considerable stockholder power, even while Barr Peat and his Pittsburgh compatriots sided with Frazer.

The judge evaluated that Willard's charges were "not frivolous," and he nullified the November 19 decision based on a procedural error in the way the vote had been taken, as indicted by the corporation's own by-laws. On this technicality, Willard won a preliminary injunction invalidating the vote for Channelair's license.

At the same time, Channelair was in the middle of making arrangements to acquire a new, lighter engine. It had already tested a model of its Channel Wing design in a wind tunnel with very satisfactory results. But then it discovered that Custer-Frazer had become, in its words, "a house divided." It learned that no license to build and sell Channel Wing aircraft would be forthcoming without Willard's express approval. Attempting to salvage the deal over the next month or two, they finally realized that in absence of an FAA Type Certificate on Five-Seven-Charlie or any other model Channel Wing, no one would be mass-producing Channel Wing parts. Channelair then offered to collaborate to obtain such certification for its design, but with Frazer involved, Willard did not trust them and would not share his certification test results of the past six years.

The relationships were completely poisoned, and the Custer-Frazer corporation quickly withered. Despite the intense early effort and its promise, it had lasted only two years. In another year, the Custer-Frazer Corporation would be declared insolvent and written off as a gigantic, costly distraction.

In March 1960, Custer-Frazer sold Five-Seven-Charlie to the Custer Channel Wing Corporation in Hagerstown for nearly $33,000, financed through a loan by Channel Wing supporters. As Heinz, the Pittsburgh contingent, and the New Yorkers abandoned the Channel Wing project, Willard somehow maintained loyalties in Montreal. Those remaining loyal included Bill Spence, President of Custer Channel Wing (Canada) as well as Bill Atrill, the Canadian pilot, who agreed to stay on in Akron a while longer.

To his credit, Atrill continued flying the CCW-5 for many months afterward, helping Willard plow through some more of the Type Certification requirements. He wrote the only Channel Wing Flight Manual ever produced while gathering other crucial data. "The channels really did work," he wrote his grandson years later. "I used to take off at 25 MPH and climb at 38, approach at 42."

One certification issue in particular needled Atrill, however, and he confided it to Willard.

"You have a severe downwash problem," he said. "The stability and control is wrong and unsafe and has to be corrected before the FAA will certify it."

Atrill had discovered at takeoff that the streams of air being pumped through the channels were rushing onto the ground and beneath the tail, forming a river of air between the tail and the runway that prevented the tail from going down. Because the nose and tail work like a seesaw, the nose rising when the tail is lowered, the nose would not come up if the tail could not go down. This made it extremely difficult to get the plane pointed toward the sky and off the ground. It also magnified the danger in maneuvers such as steep takeoffs and slow flight, which required the nose to be pitched sharply higher. It had affected Atrill during the Quantico demonstration, and he himself had had to abort one takeoff because of it.

But Willard didn't want to discuss it. First, he knew about it already. In fact, the public knew about it. Nearly ten years prior, the NACA report on the CCW-2 had noted the downward deflection of air from the channels; Ray and Ray had also reported its effects at takeoff during their CCW-5 demonstration in California; and an Aviation Week reporter had explained it in detail after Atrill's aborted takeoff at Quantico.

Second, Walt Davidson and Bill Atrill had both found a way to work around it. The trick: Don't wait too long to bring the tail down. If you slip the tail under the downwash before it becomes too concentrated, the downwash can, instead, help hold the tail down—and nose up—and make the climb that much easier. Willard would consider it a pilot training issue until he found money to otherwise address it.

The Securities and Exchange Commission

Back when the Custer Channel Wing Corporation was still the National Aircraft Corporation, and Frank Kelley was its President, the company had recorded 94 stockholders. Of these, most held only one or two shares.[10] Willard, by contrast, held 496 shares, nearly half of the 1,120 shares in circulation. The next highest shareholder was Frank with 137 shares. The next highest held 39. So, Frank always had more invested in the Channel Wing than anyone else besides Willard.

As long as the company continued, Frank provided professional photography for it, never taking credit and probably never taking pay. All the company's press release photos were taken by him. After he relinquished his early title of President, Frank continued to work in the background cultivating business relationships in the interest of the company with news media contacts, editors of periodicals, and college and university engineering department professors, among others. As a professional photographer and small-town business leader, he had a wide network of personal contacts and business associates in Hagerstown and nationwide that he built and maintained over the years.

One of Frank's more valuable contacts remained C.G. Taylor from the days of the CCW-2 demonstrations. Keeping C.G.'s pulse had gotten easier since the hub of Channel Wing operations had moved to Akron, Ohio. Ever since Taylorcraft had folded in Pennsylvania, C.G. worked as Chief Design Engineer for his family's business, Products Design, Inc., in nearby Alliance. Frank knew that C.G. still wanted to build a Channel Wing aircraft, and Frank finally found a way.

Early in 1960, as the Custer-Frazer and Channelair corporations were being dismantled, Frank learned that a fellow Hagerstown businessman—Justin Funkhouser, the President of Victor Products Corporation—had money to invest.[11] After a phone call to Alliance, and another to Oxnard, California to explore possible locations, Frank put together a plan to put C.G. to work building a Channel Wing aircraft. After he walked through his idea with Willard, Willard wanted to hug him.

"You'd best get C.G. working on it as soon as possible, Frank," he said. "He's getting to be an old man."[12]

Figure 6.4 — Command-Air model

Under the Fun-Kell (Funkhouser-Kelley) Aircraft Corporation, C.G. began to model (Figure 6.4) and then prototype the "Command-Air," a tiny aircraft with a wingspan of 20 feet, total length just 12 feet, and constructed of fiberglass over a metal frame. It sported sweptback wings like a jet and winglets (canard) in the nose. It had no tail. Directly behind the cockpit were the engine and a single channel circumscribing a propeller. It looked very futuristic (Figure 6.5).[13] Eventually the wings would be foldable, C.G. explained, but not in the prototype. He predicted it would take off in 50 feet and land at ten or fifteen miles per hour. It would be the perfect flying car, he pitched, fitting nicely into the family garage. Taylor estimated that a single plane would sell for a mere $3,000, the cost of a reasonably priced automobile in 1960. Given C.G.'s experience with manufacturing and selling aircraft, "it is believed that the Company will experience no difficulty in the securing of a Type Certificate for this design," Frank wrote in his business plan. He had learned with Willard that ultimate success depended on getting the FAA Type Certificate.[14]

Fun-Kell bought an annual license from the Custer Channel Wing Corporation for $10,000. Frank's photographic studio in Hagerstown doubled as Fun-Kell's corporate address, but C.G. built and tested a model, and then started building the prototype back in Oxnard, California, on grounds familiar to Channel Wing loyalists. That $10,000 fee constituted the Custer Corporation's sole cash income that year, but it felt good.

Then more Channel Wing fruit ripened in California. Ward Brooks, a real estate developer, probably left over from the McAllen group, placed an order for twenty CCW-5s just like Five-Seven-Charlie. This order paled next to the 100 aircraft that Custer-Frazer had committed to build, but it carried more weight because it came with a $10,000 deposit. That money Brooks put into a trust in a Hagerstown bank, to be returned in the event Willard did not deliver.

With a bonafide order for twenty aircraft, the Custer Corporation had a tangible, bite-sized goal. The Corporation would close its operations in Akron and consolidate in the hangar it rented at the Municipal Airport in Hagerstown. But the Corporation needed to acquire or borrow parts, machines, materials, and labor to build even one aircraft. And it had almost no money.

But it never had a more appropriate reason to sell stock.

Five years previously the Corporation had managed to raise $156,000 by selling stock. Now in 1960, owing to the flurry of stock splits and transfers it had borne for Custer-Frazer, the Custer Channel Wing Corporation had enough stock left to set $208,850 as a reasonable sales goal. So, on August 8, 1960 the Corporation filed Form 1-A with the Securities and Exchange Commission (SEC), informing it of its intent to offer 461,700 shares in a second public stock sale. The application included a slightly updated version of the circular it had published five years previous.

In the aftermath of the stock market crash of 1929, Congress had

Figure 6.5 — Command-Air, front (top) and side (bottom)

formed the SEC to apply rules that would protect the public from the many abuses that had contributed to the severity of the crash. The primary objectives were to ensure that potential investors were provided with material information pertaining to a particular stock and the company offering it, and to prohibit deceit, misrepresentations, and other fraud from creeping into the sale.

In order to meet its objectives, the SEC required that every company preparing to offer stock for public sale first register the sale with the SEC. The SEC would examine the registration statement to ensure that potential buyers would be given the entire truth and only the truth before they purchased. In addition, a corporation also had to file documents in every state in which its stock might be sold. For Willard's corporation, this might have included all fifty states.

The registration process can be very complicated, expensive, and time-consuming. So, in order to prevent small companies like Willard's from being discouraged at the outset, the SEC provides several possible exemptions. Most startup companies wanting to sell stock are not ready to tackle the registration process, so they turn to the exemption options for seeking funding. The Custer Channel Wing Corporation had successfully invoked the Regulation A exemption in 1954 and sought the same exemption for its 1960 sale.

Regulation A is an accommodation to small stock sales, which is to say, sales projected to generate less than a certain amount in any twelve-month period. In 2013, for example, the sales threshold was $5 million. But in 1960, the threshold sat at $300,000. The Custer Corporation proposed to sell only enough stock to generate $208,850, easily qualifying for the exemption.

Even though smaller stock sales under Regulation A may be exempt from the more extensive and demanding registration requirements, by law such sales still must be documented with the SEC, reviewed, and approved.

Given its successful stock sale in 1954 and its previous positive experience with the SEC, the directors of the Corporation felt blindsided and betrayed when on December 30, 1960, in Release 4311, the SEC refused to approve the proposed sale.

"The Securities and Exchange Commission has issued an order temporarily suspending a Regulation A exemption from registration ... with respect to a proposed stock offering by Custer Channel Wing Corporation..." The denial alone caused anguish, but its tone rang accusatory and disdainful:

> The Commission's suspension order asserts, among other things, ...that Custer's offering circular is false and misleading...
>
> The misrepresentations alleged in the Commission's order relate to information in the Custer offering circular concerning the development, manufacture and marketing of aircraft embodying a "new" wing design when, in fact,

> the design has been proposed and under development since 1940; the failure to disclose the history of such development in reasonable detail; the failure to indicate that during the fifteen year period the proposed aircraft has been under development by Custer, its predecessors and subsidiaries, sums aggregating several hundred thousand dollars were raised through the sale of securities, to disclose how such sums were expended and the reasons why a salable aircraft has not been fully developed, or to disclose Custer's previous unsuccessful efforts to market the aircraft and the fact that the aircraft was demonstrated to the military and that no interest was shown or orders taken; the failure to describe the patents pertaining to the wing, to disclose that applications filed in 1953 and 1954 with the predecessor of the Federal Aviation Agency were not completed and have since been abandoned, or to furnish an estimate of the amount required to secure FAA certification of the aircraft proposed to be manufactured; statements that the breakeven point will be reached at approximately the fifteenth aircraft produced, that the company has "firm" orders for twenty aircrafts, and that $208,850 will be enough to commence actual manufacture of aircraft to fill outstanding orders; the failure to disclose that the market price of the Class B stock is substantially lower than the public offering price; and the failure to include financial statements prepared in accordance with generally accepted accounting principles.

To add embarrassment to anger, the entire text of the response was printed in the Hagerstown newspaper. Willard and his directors cursed and swore. Why, they wondered, if the SEC had found some impropriety, had it not notified them of such in a private letter so that they could address it and revise their papers? Why had it nailed, as it were, its rejection like a broadside in the market square?

Only a temporary order, it came packaged with "an opportunity for hearing, upon request, on the question whether the suspension should be vacated or made permanent." So, practically speaking, they had the option to address each "misrepresentation" cited by the SEC and hammer out a circular the SEC would deem transparent. But they did not want to do that. Such a circular would not present a very appealing investment opportunity, and that hit at the heart of the matter. The SEC had found no issue with the

Channel Wing aircraft; the agency had faulted the company. Withering in parched clay, its biggest opportunities now past, the company desperately thirsted for money to survive. Potential investors would not be attracted to that picture.

So, the Corporation bought time: it requested a hearing, and on February 20, 1961, the SEC scheduled it for late April. In the meantime, the Corporation scrambled for alternatives. In March, for example, it applied to the Small Business Administration (SBA) for a $225,000 loan, but the SBA turned it down.

To prepare for the hearing, the SEC directed one of its investigating attorneys, Joseph Levy, to investigate the Custer Channel Wing Corporation and the Custer-Frazer company. If Levy interviewed any of the directors of Custer-Frazer (his notes are not subject to a FOIA request), he got a taste of the simmering cauldron of resentments those gentlemen had cooked up especially for Willard Custer.

Then, on April 21, the SEC inexplicably postponed the hearing, which gave the Corporation yet more time to explore other options.

Finally, in June, the Corporation withdrew its request for the hearing, and considered the matter closed. It had found another way to raise cash without having to appease this bully. Accordingly, the SEC issued the order permanently forbidding the stock sale.

But it did not suspend its investigation.

SEC v. Custer Channel Wing Corporation

While the SEC had been preparing for the hearing, the Corporation had generated a new idea for selling stock. The SEC registration rules applied only to the sale of stock owned by the Corporation in a public offering. The registration rules did not extend to the resale of stock held by a stockholder. So, the Corporation reasoned, it could resell previously purchased stock, acting as a broker, and thereby legitimately sidestep the SEC.

This, then, was the plan: Willard (and any other stockholders who chose to participate) would temporarily loan some of his stock, including valuable Class A stock, to a new trust to be named Stockholders Inventory and Manufacturing Trust. According to the Declaration of Trust, the Trust would have control of the stock for one year in which it could sell it. If the stock were not sold by the end of the period, the Trust would return it to its owner. However, if the stock were sold, it would sell for fifty-three cents per share, according to the Declaration, and the money would be "made available to Custer Channel Wing Corporation to be used by that corporation to pay the cost of completing manufacture of one or more aircraft embodying the Custer Channel Wing invention." The Trust, then, would act as a broker to sell Willard's stock and provide the proceeds to the Corporation for building aircraft. (The Declaration did not mention a Type Certificate.)

Willard would sell some of his own stock in his own corporation. He felt so desperate that he would diminish his share in ownership and long-term profits in order to provide short-term revenue.

The Corporation's legal counsel, Willard's brother, Cecil Custer, reviewed and sanctioned the plan. He named John Lawson, the Corporate Treasurer, to be Trustee.

At the same time, Frank Kelley, the current President of Fun-Kell, needed to make his second annual license fee payment of $10,000. But, as if stricken by the same drought, Fun-Kell's cash had dried up, too, and he could not pay the fee. Still the third largest shareholder of Custer Channel Wing stock, Frank proposed to follow his longstanding partner's lead. On September 6, the Corporation agreed to buy back $10,000 worth of Frank's Class A corporate stock in lieu of Fun-Kell's license fee.

Frank's stock did not go into the Trust but into the corporate treasury, and the Corporation immediately offered it for resale with Willard's. The Corporation did not notify the SEC, rationalizing again that the resale of previously purchased stock need not be registered with the agency. On this point, however, the Corporation erred, as the stock would be issuing from the corporate treasury, not simply being brokered through the Trust.

In the stockholders' meeting of September 25, and in a subsequent letter to all 2,551 corporate stockholders, Willard "fully disclosed" the existence of the Trust and its purpose. He also, quite excitedly, trumpeted a come-and-get-it clarion call. He announced the sudden availability of Class A stock "for the first time since 1947. It is very doubtful this can ever happen again. Class A Stock votes for election of Directors and Management."

Like a drought-quenching shower, sales commenced immediately and continued at a steady rate. By December 22, thirty-two investors had purchased stock from the Trust while eighty-two investors had purchased from the Kelley block, and there was more to sell.

Four buyers of Class A stock who thought they were buying from the Kelley block instead had their Class A stock fulfilled out of the Trust. All four buyers later testified that the source of the stock didn't matter to them; they only cared that the money went to the Corporation's manufacturing program. But it mattered to someone.

On Friday, November 17, SEC investigating attorney Levy and staff accountant Nicholas Anastopoulos pounded at the door of the office/workshop of the Custer Channel Wing Corporation. Their dreadful appearance sent a cold shiver down Willard's spine. They demanded records of any and all stock sales made by the Corporation since the summer. After more demands for information and conducting some interviews with the recent buyers of the stock, the SEC had enough material to approach the U.S Attorney's Office (Maryland). Then, on Friday, December 22, 1961 the SEC filed "a complaint for injunction, demanding a temporary restraining order, preliminary injunction and final judgment against the Custer Channel Wing Corporation, Willard R. Custer, and John D. Lawson, Trustee" in the U.S. District Court for the District of Maryland (Litigation Release 2166).

The Corporation, of course, acquiesced, halting the stream of stock sales until the case could be decided. It also canceled Christmas. The resulting case, Civil Action No. 13500, SEC v. Custer Channel Wing Corporation, would go to trial the following April (1962).

In the meantime, the monetary drought returned even while the stock proceeds remained somewhat protected (there was always juggling going on). Willard and his son, Curley, both gave back several months' salary to extend the withering Corporation's life a bit longer. Costs increased without remedy, and with no cash coming in, outstanding debt smothered activity. In 1961, total accumulated debt measured just shy of $180,000.

The Crash

Two months later, on Friday, Groundhog Day, February 2, 1962, the Pennsylvania rodent saw his shadow and predicted six more weeks of wintry weather. Outside the Channel Wing hangar at the Akron-Canton Airport, the afternoon temperatures remained below freezing, but winds were light, and visibility extended six miles. The ground had frozen, and so had Hagerstown pilot David Lewis and his passenger, Bill Spence, the former Custer-Frazer Corporation's partner from Montreal (Figure 6.6). With a ferry permit that expired in ten days, they boarded Five-Seven-Charlie bound for Wheeling, West Virginia on the first leg of their flight to Hagerstown. Lewis warmed himself in the knowledge that airplanes, and their engines, love crisp, cold air. Wrapped in wool caps and coats, they were stuffed inside the tiny cabin like foam rubber inside a mattress case. The plane had no heater; the flight would be numbing.

After warming the engines, Lewis got clearance from the tower and taxied onto the runway. He could still see his breath. He coaxed full power from the engines and released the brakes. Reaching his desired speed, Lewis pulled back on the yoke to lift off, but the plane did not respond. In seconds, he found himself nearing the end of the runway but still on the ground. Spence, rigid from cold, became rigid with fear. At the last second, the wheels released the pavement and the plane started to climb, but a scrub pine tree loomed directly in the path. Five-Seven-Charlie's belly scraped the top of it, jarring her occupants.

Incredulous and wondering how badly he had damaged her, Lewis immediately brought her down into an open field. Landing on the rough, frozen ground broke the gear and scarred the underbelly of the fuselage. The damage would require over 2,100 man-hours of repair.[15]

Though the accident did not harm Lewis or Spence, it did hurt the plane's safety record, not to mention the embarrassing self-contradiction of a high-lift aircraft failing to clear a tree. Had Lewis waited too long to bring the nose up? In any case, the accident was publicly reported and Willard lost boasting rights and would never be able to repeat the statement made earlier in the corporation's 1958 annual report:

> There have been seven free flight models, nineteen wind-tunnel models and three full scale Custer Channel Wing aircraft built and tested successfully for nearly 2000 hours at the cost of hundreds of thousands of dollars to assure perfection of this phenomenal aircraft for public use. Not once has there been a failure or accident with Channel Wing in flight.

The accident did not dampen Spence's enthusiasm for the aircraft, however. He continued to believe in the Channel Wing design, even to the point of applying for his own patent in 1967 for an improvement that would lessen the wing's drag.[16]

On their arrival at the Hagerstown airport without Five-Seven-Charlie, Lewis and Spence stopped in at the little white wooden hangar with "Custer Channel Wing Corp" painted in red capital letters above the doors. Laid out on the expansive floor they found "welding torches, metal hammering, overlapping wings, unfinished fuselages, burnished aluminum and airplane entrails." Rising from the middle of it all, like a phoenix from the ashes, a brand new, hand-built CCW-5 had already taken shape.[17]

By the beginning of 1962, the Custer Channel Wing Corporation had not one but two CCW-5 aircraft in the works, at different stages of manufacture and assembly in their Hagerstown hangar. Willard understood that an FAA Type Certificate had not yet been approved, but he could still build individual experimental aircraft. His drive emanated from the account in a Hagerstown bank containing Ward Brooks' $10,000, the California real estate developer's deposit on twenty such planes. At sixty-three years of age, Willard held a firm order for Channel Wing aircraft, and he would satisfy it despite all. Now he just needed the stalled stock sale to finish its course. That could only happen if he prevailed in the SEC case.[18]

After a pre-trial hearing and various intervening motions, the SEC trial opened on April 10, 1962 in Baltimore before the Honorable Harrison L. Winter. No jury was present.

The Seegmiller, Wilner & Custer law firm represented the defendants, but their attorneys were no match for the prosecutors from Washington, D.C. and the U.S. Attorney's Office in Baltimore. One early indication of this is the following statement the firm made during the run-up to the trial.

> ...the plaintiff [SEC] is, and for a long time has been, fully cognizant of...defendants' willingness and efforts voluntarily to comply with all requirements of law, but that plaintiff has, notwithstanding, repudiated its manifested intention to administratively give notice of its claims of illegal conduct (if any) by defendants to permit adjustment thereof without litigation, and in lieu thereof plaintiff has adopted a course of conduct marked by oppression, harassment, and arbitrariness toward the defendants.

The statement sounds like whining, but it also misrepresents the law enforcement mission of the SEC. The SEC does not point out a corporation's infraction of a law so that the corporation can adjust its actions and thereby avoid prosecution. Rather, the corporation's legal counsel is responsible to steer its client clear of any and all infractions when counsel is consulted. The Seegmiller law firm may have failed to do that.

Boiled down to its essence, the SEC's case alleged that the Custer Corporation had 1) sold stock without a registration statement, and 2) withheld material facts important to investors. With respect to the first charge, the SEC maintained that the law required the Corporation to register its sale of corporate stock from Frank's Class A stock that had been returned to the Corporation's treasury. With respect to the second charge, the SEC alleged that the Corporation had misrepresented stock as coming from the Corporation's treasury when, in fact, it had come from the Trust. The latter charge applied to the four buyers of Class A stock who had intended to buy from the Kelley block and then received stock from the Trust.[19]

The two sides called their witnesses and presented their evidence on April 10 and 11. Judge Winter heard final arguments on April 24, and at the close of that day he easily ruled in favor of the SEC.

But for all the barking, the final bite did not inflict much damage. Remarkably, the Judge levied no punishment, as such. The Corporation had to pay court costs ($292.44), but otherwise he levied no fine. The judge simply told them to stop selling stock without SEC approval. The Corporation did not have to refund the money it had collected in its stock sale, nor were those sales voided. Unfortunately, the sale had not run its course, and that is what hurt. So far, Frank's stock had garnered $21,000 and cut $11,000 of debt from the books, but there remained thousands of shares of stock that could have been sold.

The Corporation did not acknowledge the SEC incident by name in either its 1962 or 1963 Annual Reports. The Corporation reported that in 1961 it paid nearly $14,000 in legal and accounting fees and still owed $62,000 more, but it did not say how it had incurred them. But those numbers paled in the face of its Total Current Liabilities of $201,374.22 and Net Loss of $749,363.05 (1962 Annual Report). At the end of 1962 the Corporation still owed $96,661.54 in attorneys' fees, nearly three times what the ill-fated stock sale had generated.[20] The stock sale had turned out to be quicksand, and the harder the corporate officers had flailed, the faster they had sunk.

To worsen matters, Fun-Kell's money dried up entirely that summer of 1962, and while C.G. had succeeded in building a prototype, ground testing it, and registering it with the FAA, the money had evaporated before he could even attempt a takeoff.[21] For the time being, he would stay in California and work on other things, in the hope that Frank could bring in some more funds.

Despite his Corporation's failure in court and its record debt, Willard somehow remained positive. Good things were still happening. At the annual meeting of stockholders held August 20, 1962 at the Hagerstown Municipal Airport hangar, Willard showed attendees the first production model CCW-5, nearly complete. He also showed a second production model, already begun, but he impressed on his audience that the first one had to be "completed and certified before any major production program could be launched. For this reason our major effort would be concentrated on certification during the coming months."

Secondly, the Defense Department had followed through on commitments made after the Quantico demonstration. In his 1963 Letter to Stockholders, Willard reported:

> Proposals presented to the military developed a request for a demonstration. It will please you to know the military not only saw a demonstration of our CCW-5 on December 12, 1963, but three top military fighter pilots actually rung out the CCW-5 for several hours, putting it through maneuvers far in excess of anything our pilots ever did. This proved the CCW-5 would meet the performance for which it was designed. We have reason to believe this demonstration will develop further interest from the military.

Perhaps the best news he saved for last:

> It was reported by the Corporation's legal counsel that all litigation in which the Corporation had been engaged had been terminated.... The Company is not now involved in any litigation, and there is no reason to expect that it will be in the foreseeable future.

In great attempts it is glorious even to fail.

– Cassius Longinus

VII The Production Model 1964 - 1973

I made up my mind, if the good Lord lets me live, I'ma gonna show 'em. When I started out and built the first models, I was ridiculed so much, everyone made fun of me so much, I was called a crackpot, and even my family thought I wasn't right... Well, I learned this the hard way, ...this is an honor, because all the people who left their mark on the face of the earth have been called crazy.[1]

- Willard Custer

Saturday, July 4, 1964 dawned gorgeous in Hagerstown, Maryland, as if made-to-order: mid-80s, blue skies, and low humidity. The Hagerstown Municipal Airport offered no shade except in the hangars, and on hotter days the tar joining the concrete slabs of the hangar aprons would bubble. But that day, dressed in their creased uniforms, flowery dresses, conservative suits and narrow ties, four thousand people gathered in the pleasant open air. More than two hundred specially invited guests had accepted invitations, including Oliver Pettit, an FAA inspector, and Congressman A. Sidney Hurlong, Jr. from Leesburg, Florida. The ambassadors of Tunisia and the Republic of Dahomey had also accepted, along with guests from the embassies of "Mauritania, the Malaysian Republic, India, Burma, France, Great Britain, Korea, and the Central African Republic," and nine more countries. Representatives of the news media had also been invited and were bussed in. From loudspeakers the air carried the sounds of Earl Barr's country music band and two other musical groups, with local disc jockey Dale Eash as master of ceremonies. Organized by the Hagerstown Chamber of Commerce, the jubilant and patriotic 1960s holiday of carnival rides and snow cones, balloons and hotdogs, games, contests, and a miniature train, all converged to celebrate American independence, a remarkable airplane, and the local man who had created it. They called it, "Custer Channel Wing Day" (Figure 7.1).[2]

Figure 7.1 — Promotional flyer for Custer Channel Wing Day

At 2:00 p.m. the formal ceremonies began. The doors of the little wooden hangar emblazoned "Custer Channel Wing Corp" slid open, and two aircraft were rolled into the blazing sun and presented to the crowd. The planes resembled a father and son: the older plane, Five-Seven-Charlie, coated white with a red cowl around the windshield; the newer one dressed in fresh aluminum, polished to sheen but unpainted, waiting for your-company's-livery to be applied. This one had been long awaited, as if born after a ten-year gestation. Now, this, the first production model CCW-5, represented the first of a fleet of twenty soon to materialize if all went according to plan.

It had been "built from [the] ground up in Hagerstown, using employees temporarily furloughed from [the] nearby Fairchild plant. Workmanship is excellent," Walt Boyne, the aircraft historian and writer, wrote. It had orthodox all-metal construction except for some fiberglass components in the channels. The twin engines were 260-horsepower Continental O-470s with fuel injection, driving seven-foot propellers (Figure 7.2).[3]

The production model appeared, at first glance, identical to the prototype. They had both been built using the same templates and drill jigs. But a closer look revealed that the newer plane had a sleeker profile facing the wind. The channel joined the fuselage more elegantly, the channels themselves looked thinner (only twelve inches), the engine casings (na-

Figure 7.2 — The two CCW-5s

celles) slimmer, and the engine braces more simplified. These slight changes did not alter the plane's performance or flying characteristics very much, though. Boyne called its interior "Spartan" compared to its predecessor's.

But Willard considered it the most beautiful aircraft he had ever seen.

The new plane stopped rolling and the wheels were chucked in preparation for its christening. Willard, in hat and suit, handed the bottle of champagne to his wife, Lula, who intended to break it over the plane's nose. Dressed in a powder pink suit and hat, and much smaller than Willard, she could not, however, reach the nose. Someone briskly supplied a nearby concrete block for her to step up on, and with her husband holding her purse, she christened the plane to general applause and laughter (Figure 7.3).[4]

Figure 7.3 — Lula christens the first production model CCW-5 while Sen. Matthias looks on.

Led by Willard and Lula, the celebrants then proceeded a short distance where, flanking the airport's primary runway, a portable dais squatted, skirted in red and white drapery and presenting American flags. Willard and Lula took their seats, joining Evan Crossley, president of the Washington

County Board of County Commissioners, and Joseph Firey, president of the Hagerstown Chamber of Commerce. Master of Ceremonies Eash silenced the crowd, and Crossley and Firey, in turn at the microphone, each congratulated Willard on his remarkable achievement.

The speakers announced that today the town celebrated the 25th anniversary of the Custer corporation, formed to protect and promote its Channel Wing design. They likely also mentioned that he had also had a hand in the formation of three other corporations, in McAllen, Texas, New York City, and Montreal. In its history, Willard's corporation had employed several professional pilots and contracted with teams of engineers in at least three states, while paying accountants, attorneys, and several full-time office personnel.

They also mentioned that, to date, Willard had been granted no fewer than nineteen patents. (Seven more were yet to follow.)

Continuing, they also pointed out that Willard's aircraft had become famous. More than five magazines had featured his creations so far, including *Popular Mechanics*, which had pictured "the Bumblebee" on its cover and would picture the prototype CCW-5 on the cover of its October issue in four more months. Countless newspapers and periodicals around the world had run stories and shown pictures of the Channel Wing's feats over the past sixteen years. His aircraft models had flown in one of the country's first university wind tunnels and been tested in the military's wind tunnel at Wright Field. The full-size CCW-2 had been tested in the nation's largest wind tunnel at Langley Air Force Base.

Finally, the speakers reminded the audience that Willard had maintained an exhausting promotional tour over twenty-five years, demonstrating his aircraft to the three services at military airfields in Louisiana, Oklahoma, Maryland and Virginia, to commercial audiences in California, Texas, Ohio and Pennsylvania, and promoting it to other potential civil and military investors and buyers from Mexico, Canada, Japan, New Zealand and Australia. Many of these previous audiences were represented in today's gathering.

Following these congratulations, Congressman "Mac" Mathias, Jr., of the sixth Maryland district and a future career senator, stood to address the celebrants. Mathias lauded Willard's perseverance and tenacity of spirit. In this respect, Willard showed a kinship with the English, French, Swiss and Scottish pioneers who had originally settled in Washington County. Mathias credited a similar courage and indomitable will in the face of challenge, discouragement, setbacks, and opposition to enable Willard to finally complete his first production Channel Wing aircraft. He hailed his success as "a fitting example of independent, persevering aeronautical pioneering."[5]

Matthias then invited Willard to the microphone. After expressing his sincere gratitude, he announced that the shiny, new CCW-5 would immediately start a test program aiming to obtain an Approved Type Certificate. The Corporation had hired DeVore Aviation, he said, a New York professional engineering firm recommended by the FAA, to complete the project. DeVore had already submitted "an engineering report on wing air pressure studies" to the FAA and was currently preparing "a basic loads" report.[6]

Willard also announced that he expected that the FAA "type inspection authorization will be received in about nine months and the aircraft will be certificated in a year." Whether DeVore agreed with that estimate, the FAA certainly did not. According to *Aviation Week and Space Technology*, the FAA estimated that about 95% of the certification work had yet to be done. The certification process had restarted from the beginning with the new aircraft.

Finally, Willard described his manufacturing goal, to have eight more production CCW-5s "completed, or nearly completed" by the time the Type Certification testing completed. The anticipation of victory thrilled the audience.

After Willard's remarks, the air show began. As he had for the Quantico demonstration, Willard explained the principles of the "speed of air," and how these aircraft were uniquely designed to employ them. Then he put them through their paces. Willard's two prize Channel Wing aircraft—first the prototype, followed by the trophy production model—roared, then leaped into the air after stunningly short take-offs. Circling and re-approaching the dais one at a time, with tails low and noses pointed high, they floated by at extraordinarily slow speeds. They capped the brief demonstration with what seemed a salute to their maker as they swept low in formation before him, acknowledging his pleased smile, finally circling and landing in uncommonly cropped stopping distances.

The audience applauded and cheered, and then the music resumed. Guests slowly headed to their cars. Some friends offered private congratulations to the triumphant couple seated on the grandstand. Willard graciously accepted their well wishes and basked in the day's glory (Figure 7.4).

The Last Stand

Willard and his corporate stockholders sincerely believed that, in the Channel Wing, they had the most important discovery in aviation since the Wright Brothers' original flight. Contrary to their naïve expectations,

Figure 7.4 — Willard, Lula (far right), and their four children (L-R, Reed, Vivian, Helen, Curley) on Custer Channel Wing Day

however, aircraft manufacturers had not caught their vision, but had reacted with resistance and competitive self-interest. C.G. Taylor had been the singular exception. The Department of Defense (DoD) had shown some interest, but it viewed the Channel Wing as a solution looking for a problem, and the DoD couldn't identify a mission that a Channel Wing could uniquely serve. In Willard's view, the aircraft companies and the DoD continued to encourage each other's doubts and resistance. Should one acquiesce and eventually buy into the Channel Wing concept, he believed the other would likely capitulate, as well.

So Willard, despite deep misgivings, had been forced to start his own manufacturing effort. Some specialized markets had opened with potential customers interested in buying Channel Wing aircraft, if they could be produced. Now, at 65 years, he could still make his mark in the industry, but not without that cursed Type Certificate.

Willard had a complex relationship with the federal government. He viewed the SEC as actively hostile, bent on destroying the Corporation's ability to raise money, every company's lifeblood. By contrast, the FAA seemed passively immovable, obese in its list of design and test requirements, which only grew over the years, as it continued to run up costs and further delay the Type Certificate. If in the future there were no change for the positive in both relationships, either one alone would bankrupt the Corporation and kill the Channel Wing. The clock ticked against Willard and his stockholders.

On "Custer Channel Wing Day," Willard declared that the Corporation had arrived on the verge of success despite a continuing list of challenges. The debut of the CCW-5 production model proved that the Corporation could produce a marketable Channel Wing, and it still had a firm order for nineteen more. The Corporation just needed more time and money to bring the effort home. With rosy projections and sales hyperbole, Willard announced that he expected it would take only one year more to finally get the Type Certificate, and in the meantime, the Corporation would be building perhaps eight more.

But Willard could not explain how the Corporation intended to pay for it.

Few in the audience knew that the Corporation sat hundreds of thousands of dollars in debt, with no cash inflow, and therefore unable to qualify for a business loan from any responsible financial institution. It had no property or product to sell or sell off. Personal loans were received occasionally, but they were dwindling, now few and small. Since Heinz' withdrawal, no more capital investors offered partnerships. And a federal judge had forbidden the Corporation to sell any more corporate stock. Indeed, no funding sources remained.

But, as the dignitaries had pointed out in their speeches, Willard would never give up, especially when success smelled so close. So, with no other options, and in their desperate push to realize their dream, Willard and his associates had managed to raise some money the only way conceivable: they had returned to selling the donated corporate stock.

Between May 25, 1962 (the date of the federal injunctive order against them) and February 16, 1965, the Corporation sold an additional 1,579,590 shares of Class B common stock through the Corporation Trust Company, the corporation's transfer agent, without notifying the SEC. The stock had been previously owned, their lawyers reasoned, and the sale remained private, so it could not be subject to SEC oversight. The Corporation later maintained that they had told each buyer that the stock sale sat outside the SEC's purview, and they offered whatever corporate financial documents the buyer cared to see.

By February 1965, the Corporation raised $400,000 through this activity, most of it having been collected prior to "Custer Channel Wing Day." So, Willard had that money in the bank when he foresaw the production of eight more aircraft by the same time the next year. For him, it was a vision; and if enough people and the right people believed in it, it would come true. Sitting on the platform on that sunny Fourth of July afternoon, amid the crowd and the cheers, the realization of his dream seemed only inches away.

But away from the platform and the well-wishers, Willard also knew deep inside that he was engaging in some very risky business. He had to impress the FAA and advance the Type Certification effort. At the same time, he had to stay one step ahead of the SEC until the FAA work completed. He was desperately racing against time to avoid a total shutdown.

Nine short months later, after dinner on Tuesday evening, April 20, 1965, Willard and Lula were home watching *The Huntley-Brinkley Report* on WRC-TV when they heard a rap on the front door. They looked at each other; neither one expected visitors.[7]

"Well, I wonder who that is," Willard said, as he rose stiffly from his deep, worn, turquoise armchair. Before he reached the door, another rap insisted he hurry, and Willard heard the announcement through the glass panes that flanked the door.

"Open up! United States Marshal!"

The sun hung low in the sky, and the large cedars that stood like pallbearers around the front porch had already shrouded the entrance in darkness. Willard flipped on the porch light and cracked open the door, a little unnerved by what he had just heard.

A young man in a black business suit stood there as if he owned the place, a step back from the threshold.

"Willard Custer?" the deputy marshal demanded, presenting his silver star badge. Willard opened the door more widely, so he could inspect it.

"Yes, sir?" Willard said, not yet knowing what to think. Lula arrived at his side.

The thirty-year-old deputy marshal had not expected to find an older man and his tiny, elderly wife. Wrapped in a knitted afghan for protection against the twilight air, Lula appeared breakable.

On seeing her, the deputy returned his badge to his black wool coat pocket and removed his hat. Then he asked, in a more neighborly tone, "May I come in?"

As Willard opened the door, the marshal stepped just inside the door. "Ma'am," he said, nodding to Lula, "I'm sorry to disturb your evening."

As the door closed behind him, he pulled a trifold paper from another coat pocket, and handed it to Willard. "Mr. Custer, what you have there is a warrant for your arrest," he politely explained, noticing how the paper quivered in Willard's hands as he slowly unfolded it.

They were older than his own mother and father, the officer estimated. This felt uncomfortable to him in many ways, but he had his duty.

Willard, hardly able to keep his thoughts together, realized he read the words "criminal contempt of an injunctive order." *Criminal?*

"There must be some mistake," Willard inquired with lungs that wouldn't work. He searched the marshal's face as he returned the paper.

"Probably so, Mr. Custer," agreed the officer sincerely, stuffing the warrant into his pocket, "but we can't straighten that out here. Sir, you will have to come downtown with me. Please get your coat and hat; it's chilly."

As Willard obediently reached inside the coat closet, he instructed Lula, "Honey, call Helen."

Nightmares

Helen arranged for the $2,000 bail and Willard returned home. Two weeks later, Willard and the Corporation filed responses denying that they had willfully disobeyed the injunctive order. Within mere weeks they stood again before Judge Winter, the same judge who had tried them three years earlier.[8] The Seegmiller firm again represented the Corporation, but without Cecil Custer who had died of a heart attack in October 1963. Leon Pierson personally represented Willard.

"The trial, which lasted approximately four weeks, began on June 15, 1965 and with an intervening court recess did not terminate until September 14, 1965," according to the SEC's Litigation Release No. 3369.[9] The defendants were found guilty as charged. The Hagerstown newspaper published all the details. In one brief year Willard had plummeted from his proudest day to his personal ruin.

This time Judge Winter punished the Corporation with a fine of $5,000 and half the government's costs of litigation (half the costs amounted to $3,512.25). Then he sentenced Willard to 183 days (just over six months) in prison and fined him the other half of the government's litigation costs. He set Willard's sentence to begin on Monday, November 22, 1965, just before the Thanksgiving holiday.

Stunned into inaction, most of the officers of the Corporation simply tried to coax feeling back into their hands and feet. Willard, however, thought of nothing but prison and Lula.

On the morning of November 22, Willard's attorney filed a motion for a new trial or a reduction of sentence. Until Judge Winter could consider the motion (he would shortly deny it), he delayed the start of Willard's sentence to Monday, December 20. Mr. Pierson then filed Willard's appeal in the U.S. Court of Appeals in downtown Richmond, Virginia on December 17, three days before the new sentence start date. Willard had to pay another $2,000 bail, but he remained free pending the outcome of the appeal.

The impending prison sentence loomed over Willard like a recurring nightmare. His appeal was finally heard February 8, 1967, nearly two years after his arrest. The transcript of the case reveals two important facts about the Corporation's illegal stock sale that were not clearly marked in the initial trial. First, the Corporation claimed that it had relied on the opinion of its lawyer, Cecil Custer, that the sale of the stock did not constitute a public offering. The court would not let them blame Cecil, however. It found that "the defendants did not in good faith rely upon the advice of counsel, for they sold and issued stock to persons far in excess of the maximum number cautiously advised by the lawyer."

The second important fact revealed in the appellate court's review concerned character. The SEC admitted that Willard and his officers had acted solely out of desperation to save their business from failing, not out of any intent to defraud the public. Any acts of deceit, then, were intended to deceive the SEC, not the stock purchasers. It was an important but fine line to draw, on moral grounds, but it did not constitute a defense. So, the appeals court upheld Judge Winter's decision.

Exhausting every option, Willard finally appealed all the way to the U.S. Supreme Court. In May 1967, the Solicitor General, Thurgood Marshall, studied his request but found nothing warranting Supreme Court review.

Willard's dread of prison time again invaded his sleep.

Less than two weeks later, on January 3, 1968, Willard's attorney filed a motion to reduce Willard's sentence, based on Willard's failing health. He was 68. Judge Winter ordered him to present himself to a court-appointed physician for examination four days later. The doctor determined that he did indeed suffer ill health, and based on the doctor's report, Judge Winter granted the motion.

To Willard and Lula's huge relief, Judge Winter cancelled the prison time, but only on three conditions. First, Willard would be on probation for two years, but he would not have to report to a parole officer. Second, the judge forbad him from having any part in fund-raising activity to benefit the Corporation. These two conditions Willard could willingly satisfy. But the third condition, though not unexpected, crushed his soul:

> That the defendant will resign as an Officer and Director of the defendant, Custer Channel Wing Corporation, will not accept any office in the corporate defendant or any subsidiary or affiliate thereof, and will discontinue any role in the management of the defendant corporation, except that the defendant may be employed as an engineering consultant to provide consulting engineering services from time to time, only.

Willard had no choice but to comply.

The Blick Theory

Before January ended, Frank Kelley received a letter from C.G. Taylor,[10] still in California, still waiting for more Fun-Kell funding. C.G. confided, in part,

> ...Last night I didn't get to sleep till after 4 o'clock in the morning just going over everything. My wife has been after me to quit working so hard at the shop and settle down to a quieter life. So I decided last night to start selling everything and go along with her. It will probably take 3 months or more to liquidate and everything must go...Something has to be done about the little airplane as I positively will have no place to move it to...Hope things are moving right along on Custer project...

Frank replied three weeks later.

> ...To put your mind at rest that I will be able to swing the finance end of the project, I am the third largest stockholder in the Custer company...I have an order to sell several thousand dollars' worth...I thought I could get out before now but with remodeling the studio and my mother not at all good, I thought it would wait a month or so. CG, I am going to have plenty and I want to spend it on our projects. So it is good to know you will now be free to go back to work on our baby. Keep me posted on your progress and where you want to do this at. I would guess the Oxnard area.

Whether or not Frank ever cashed out, their "baby" would never fly. Two years later, thoroughly frustrated, C.G. angrily told the FAA that he maintained the futuristic Channel Wing aircraft for several years afterward while waiting for Frank to supply some more money, "but when it became very evident that no further money would be available to complete it the airplane was junked."

With Willard's removal in 1968, the Corporation did not fold, but it did gradually reorganize, introducing some new names while retaining old stalwarts. The new Chairman of the Board was Jack Kough, PhD. Dr. Kough was an educator and Vice President of Science Research Associates, Inc. in Chicago through which he had developed considerable expertise in developing and publishing programs for gifted children.[11] He owned a ranch in California, and he had cash for the Corporation. He insisted that some of it go to Willard, and that Willard be retained as an engineering consultant. Willard responded in writing:

> You will recall I indicated to you what would happen if I told Mrs. Custer of your most gracious suggestion. In view of this and the fact you wrote me a letter insisting that I take $12,000.00 of the funds advanced to the Corporation by you and use it to pay off notes Mrs. Custer and I owe, plus a salary of $400.00 per week, I just handed the letter to her when I came home. As she read it, you should have seen the expression on her face. Then she reread it, turned it over backwards, then read it again. She looked up at me with the most astounded expression I have ever seen on her face. It was just impossible in her mind that a human-being lived like Dr. Jack Kough who was interested in our welfare first and the Corporation next.

Eventually, Judge Winter granted Willard permission to sell 34,000 of his shares of Class A Channel Wing stock through a broker. Kough bought it as soon as it hit the market and thereby bought a good portion of Willard's ownership in the company.

In another leadership change, Sam Stoner, a member of the Board of Directors since 1962, was elected the Executive Vice President. Stoner lived in Enid, Oklahoma where he had been mayor and president of the Chamber of Commerce, and he had many business contacts there. Stoner had met Willard and Five-Seven-Charlie in August 1957 when Willard had spent three weeks in Oklahoma, promoting the Channel Wing and demonstrating the CCW-5 as part of its transcontinental tour from Ventura, California to

Akron, Ohio.

Now, under Stoner's leadership, the corporate address and activity hub were moved to Enid for a fresh start. Stoner had a new location, but he maintained the old strategy: produce a study to prove legitimacy; work for FAA certification; demonstrate to the military.

In Enid, Stoner commissioned a study into the Channel Wing's aerodynamic properties by the University of Oklahoma's Department of Aerospace and Mechanical Engineering. Dr. Edward F. Blick, a Professor of Aeronautics and a STOL expert, led the study.

Blick approached the Channel Wing differently than his predecessors had. Building on past studies, including Dr. Crook's wind tunnel tests and the NACA results, Blick wanted to develop a mathematical model that the Corporation could use to predict the amount of lift advantage a channel wing would provide if added to any given fixed wing aircraft. Blick would avoid repeating past tests; he wanted to advance them to the next step. In fact, he had something that none of his predecessors had had at their disposal during all the years of prior research. Blick had a computer.

"The experimental data indicate that the channel wing is a surprisingly

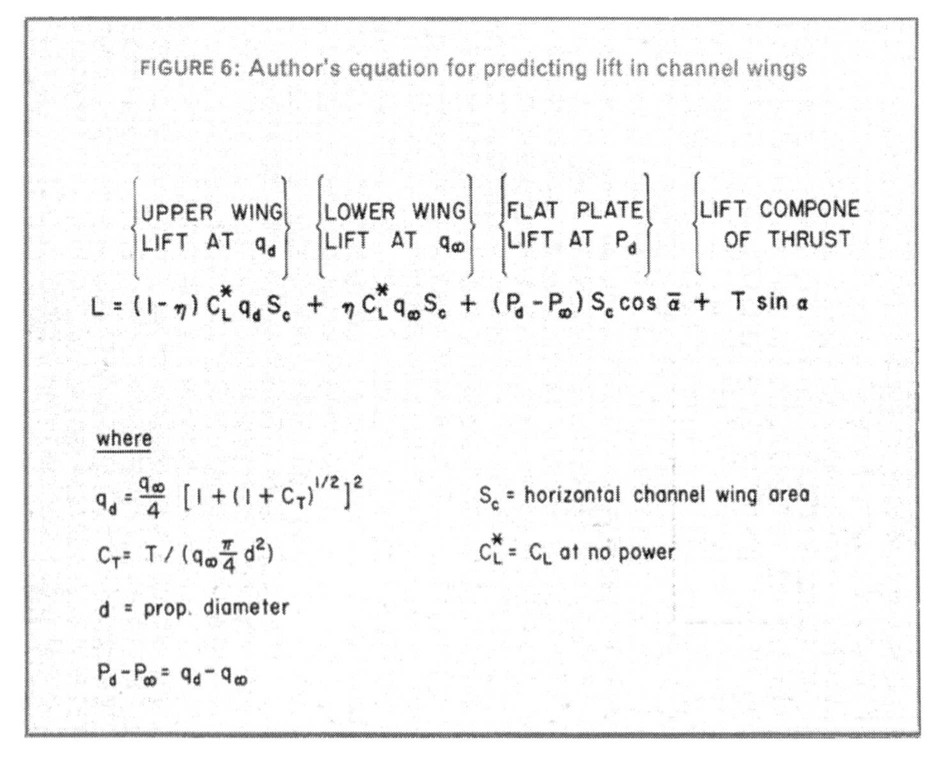

Figure 7.5 — The Blick Theory

efficient lift generator. The largest channel wing efficiency measured, 13.7 lb/hp, is in the same range as those generated by helicopters," he stated in the *Journal of Aeronautics*. With this performance record, why had the Channel Wing not been adopted commercially or otherwise? "Although several full-scale channel wing aircraft have flown and demonstrated STOL capabilities since 1943, there has been a dearth of theoretical and experimental research applied toward this concept. Without doubt, this lack of satisfactory theory and design formulas for predicting lift on channel wings seriously impeded the development of channel wing aircraft," Blick proposed. So, he set out to provide what was lacking.[12]

By December 1969, Blick had developed what he called the "Blick theory." He could use his formula on existing aircraft with known takeoff characteristics to predict how lift would increase if a channel wing were incorporated in the design. His formula, as published in *Shell Aviation News*, appears in Figure 7.5.

During the same period, the Corporation under Stoner continued working with DeVore Aviation to complete Type Certification testing. Headquartered in New York, DeVore took control of the newer CCW-5 (Five-Seven-Charlie was retired by then) at Linden Airport in Linden, New

Figure 7.6 — Willard (left) and Dr. Blick (right) confer in Linden, NJ

Jersey, in 1968 (Figure 7.6).[13] Their engineers had originally done some initial flight-testing of the aircraft and made some "theoretical determinations" in June 1964. That data had provided some material for Blick's study. Now, in 1968, the firm painted the aircraft smartly in red, white and blue livery, and put it into the sky. Prior to 1968, the plane had logged only 27 hours of flight time. But in their first year, DeVore nearly doubled that amount.

FAA records show that DeVore removed the CCW-5 from service for all of 1969, as it "was being modified to meet F.A.A. certification requirements."[14] The modification raised the tail wings (horizontal stabilizers) several inches higher on the tail fin (vertical stabilizer).

Five-Seven-Charlie had exhibited a pronounced downward backwash from the channels that could make high-performance takeoffs tricky if the pilot wasn't properly trained. Pilot Bill Atrill had predicted that the FAA would never grant a Type Certification until Willard remedied the problem. According to Dr. Blick, this is not the reason DeVore raised the horizontal stabilizers on the production model. He said they did it "to achieve a smoother flight during cruise." However, their higher position did affect the CCW-5's takeoff. Whether it solved the downwash issue, it hurt performance, according to Blick, lengthening the takeoff and landing distances by seventy percent (Figure 7.7).[15] These are the types of changes that Willard had always resisted.

By 1970, as the Channel Wing camp's funds lessened, the frustration with the FAA increased. Blick reported that FAA "tests were admittedly negative" due, in his opinion, to "some improper conclusions and erroneous statistics used by the government testers."

In the end, the larger, philosophical differences would become the insurmountable ones. FAA guidelines, geared toward conventional airplanes, could not be applied to the radically unconventional Channel Wing. For instance, a Channel Wing "will fly without wings as known on conventional airplanes but this aircraft had to have wings just because the FAA says all airplanes must have wings," Bruce Wallace, one of the directors, complained. The Corporation and the FAA both claimed willingness to compromise on the issues, but when design compromises would change a Channel Wing into something other than a Channel Wing, compromise had no room. The more magnanimous members in the Channel Wing camp believed the officials would give their designs more latitude as they came to understand them but educating the FAA would only lengthen the already costly, laborious, time-consuming process.[16]

"This airplane may go down in history as the one which has suffered the most in trying to hurdle the stringent requirements of certification," a

Figure 7.7 — CCW-5 showing horizontal stabilizers at their raised position

Canadian reporter had written back in 1967. Bill Spence of Custer Channel Wing (Canada) estimated then that "almost a total of $1,000,000 is required to conduct numerous tests with versions and sections of the channel wing before certification would be granted."[17]

One million dollars.

On Monday, March 16, 1970, the Custer Channel Wing Corporation re-introduced the modified CCW-5 to the public.[18] In a demonstration at Teterboro Airport in New Jersey, Curley Custer, Willard's oldest son, twice demonstrated its short takeoff and landing talents (less than 200 feet) and its slow-flying ability. Onlookers reported in *The New York Times* included "Dr. Andrew W. Cordier, president of Columbia University, who is a director of the Custer Channel Wing Corporation." Willard attended, too, described as a 70-year-old "technical adviser and stockholder of the company" who had built four Channel Wing airplanes in his lifetime. Edward Hudson, the *New York Times* reporter who described the demonstration, reported that the Corporation hoped to "build and sell commercial versions as STOL (short takeoff and landing) airliners serving airports the size of a football field in downtown areas."

Hudson also reported that the University of Oklahoma (Dr. Blick) had recently completed its two-year research contract to study the aircraft. Hudson related Blick's description of the Channel Wing as "a 'very viable' concept with 'tremendous potential.'" For example, Blick offered, "a Lockheed Electra turboprop airliner, which carries about 100 passengers, could take off in 252 feet if fitted with channels around each of its four engines. The Electra's takeoff run is regularly 2,620 feet," he said. Employing his theory, then, Blick predicted the use of channel wings could shorten the Electra's takeoff run by an astonishing ninety percent.

The Teterboro demonstration did not mark the beginning of the next phase of FAA testing, but the end of it. The FAA Type Certification tests on the airplane were not complete, and they would never resume.

Dr. Kough did not have a million dollars. And what he had put into the coffers, again, had been spent.

Thirty days after the demonstration (now April 1970), the CCW-5 flew to Enid, Oklahoma. Kerr Aviation Services hangared the aircraft at the Wiley Post Airport. In exchange for his hangar space, Kerr Aviation's owner, Robert, accepted a seat on the Corporation's Board of Directors. Kerr and Stoner then began lobbying Oklahoma's Congressman Tom Steed to push the chairman of the House Armed Services Committee L. Mendel Rivers to arrange "a good honest evaluation of the aircraft," as they called it. They would angle one last time for that elusive military contract.

Their lobbying paid off. They arranged to demonstrate and evaluate the aircraft before the end of the summer at the tiny Cimmaron Field (now Clarence E. Page Airport) in Yukon, Oklahoma. The Air Force contracted with Dr. Barnes McCormick, head of the Department of Aerospace Engineering at Pennsylvania State University and an aerospace consultant, to conduct the test.

The CCW-5 performed four takeoffs and landings for which McCormick measured the ground distances using a tape measure and clocked the acceleration and stopping times using a stopwatch. The report went to the Air Force, which never released it to the public. But McCormick's impressions were positive enough that he published a stock photo of Five-Seven-Charlie with his general analysis of the CCW-5 in a university textbook he later wrote, entitled *Aerodynamics, Aeronautics and Flight Mechanics.* "Your author viewed critically the slow speed performance of the aircraft... and was impressed by its STOL performance," he wrote. His only criticism: it was loud.[19]

Dr. Blick, however, wrung his hands over the demonstration results. According to one newspaper, the CCW-5's takeoff and landing distances averaged 405 feet. A comparable, conventional airplane would have needed another 1,000 feet, but 405 measured more than twice as long as the distances demonstrated at Teterboro. Blick explained that the longer distances were due to the summer's high temperatures, which he said diminished the engine's power output and lowered air density, thus reducing lift.[20]

Following Dr. McCormick's evaluation, the Corporation crafted and submitted to the Air Force an unsolicited proposal for development and production of a Channel Wing aircraft for military uses. The estimated cost? $19 million. The Air Force did not even nibble.

By the following summer (1971), Kerr had lost confidence in the venture. His company had been maintaining the aircraft for free since it arrived in Oklahoma. Even though he still held a seat on the Corporation's Board of Directors, he'd become annoyed at the lost hangar revenue, so he obtained a $3,200 lien on the aircraft for nonpayment of the past year's expenses. That in turn annoyed another board member, Bruce Wallace, who willingly loaned the Corporation the needed amount to pay off the lien, leveraging the CCW 5 itself as collateral. The aircraft had become a burden and members of the Corporation were divided in their feelings about it.

But one thing they were united on. After thirty years of valiantly cheating death, the bankrupt Custer Channel Wing Corporation had no more tricks or lifelines. It had reached its end, and for all practical purposes, it shut its doors.

The many lawsuits for millions of dollars would gradually end without satisfaction.

The stock certificates would wane in value except as collector's items to be sold on eBay.

But the emotions would live on for generations.

And so would the idea: *it's the speed of the air!*

Flying Home

Retirement for Willard tasted bittersweet. As much as he grieved the loss of his company and his aircraft, he needed the rest. No longer able to steer a company in crisis, he worked only part-time, even if he still worried and lobbied full-time.

With his corporation and his airplane removed to Oklahoma in 1970, Willard tried to market what remained without running afoul of Judge Winter. He was no longer on probation. And he still had a few active patents in his own name and the technical drawings from the prototype CCW-5. By 1973 he formed a new corporation, W.R. Custer Channeled Aircraft, Inc. and he appointed Rev. Robert Whitehead its president.

Bob Whitehead had befriended Willard back when they had met in the Harrisburg, Pennsylvania area in the 1960s, near the time the first production CCW-5 had been unveiled at "Custer Channel Wing Day." A former Marine during the Korean War and an avid aviator (he had helped to establish the Gettysburg chapter of the Experimental Aircraft Association), Whitehead pastored a Lutheran church when he had met Willard. He had no personal funds to invest in the Channel Wing, but he knew a lot of people and believed in the aircraft and the opportunity. He had selflessly offered to help Willard in any way he could.

Via this new corporation, and with Bob's help, Willard offered for sale the complete drawings of the prototype CCW-5. They were "full sized production drawings containing the most exact detail," the sales brochure said, "executed by professional draftsmen under the supervision of a certified FAA appointed" designer and containing the "pertinent information for fabrication and/or assembly."[21]

But Willard would not sell them to just anyone. The buyer must be intent on building a Channel Wing aircraft. Reading and interpreting the drawings would "require a person with skill and experience," and using them to build an aircraft was "beyond the range of the minimum tools, but it is not an impossible project for an experienced, tenacious builder." The Corporation brochure even offered Willard's direct assistance, but on one condition: "Like all projects a complete commitment is essential."

No one could ever show his level of commitment, and he received no suitable offers.

In 1975, a parishioner in Bob Whitehead's congregation donated $10,000 to Willard, which Willard then used to employ Bob full-time for one year. About the same time, Bob became aware that the Custer Channel Wing Corporation had ceased maintaining the CCW-5 in Oklahoma and had let it fall into disrepair. Almost twelve years old, the nose gear had been damaged and remained locked in the gear-down position. What else needed fixing or replacing? Whitehead would go find out.

He made a few phone calls. One of his local Pennsylvania acquaintances—a dentist and experimental pilot—agreed to house the aircraft on his personal property if it could be retrieved. Bruce Wallace, still holding the lien on the plane, forgave it and released the plane to the Corporation, which, in turn, to the relief of all, released it to Whitehead.

And that is how, in September, Bob found himself in Oklahoma, from where, on Saturday, the twenty-seventh, he wrote to his wife, Faye. "The CCW-5 is in pretty good shape after being abused for 4 years. Have the

Figure 7.8 — Rev. Bob Whitehead

engines tuned up, props are excellent. After we get the ferry tanks installed we will be ready."[22]

The aircraft had used a bladder tank for fuel, located between the spars in the fuselage, but, Whitehead wrote, "it would never have held up on the flight home," so he removed it. Instead, he arranged to have larger steel tanks installed, capable of holding "almost 90 gallons, so we'll have plenty between 2 hour hops."

On Monday he discussed the arrangement with the FAA field officers. "They were of great help although I thought the separate shutoff valves were a bit much—but they could have refused completely," he wrote. "Willard is pleased with the way I've handled things and is thankful that I could take care of everything."

A month later, however, he still had not arrived home. From Rolla, Missouri, on October 25 (Saturday), he wrote Faye. "This has been a strange week! I truly thought that I would have been home by this time, however this should be the *last* problem!"

The problem concerned the propellers.

"Am truly pleased with the ease of the plane to fly," he wrote, relaying some good news. This was Bob's first and only opportunity to fly a Channel Wing aircraft. "The controls are so well balanced...and it trims up for 'hands off' flying—if I can get the props fixed—it will be just great." And Bob fixed them.

Despite the personal risks, he flew the crippled CCW-5 back to Pennsylvania. He landed safely at Reigle Airport, five miles from Hershey, and quietly stashed the plane in the dentist's barn at an address unknown even to Willard.

The first production Channel Wing aircraft—one that had initially been christened before 4,000 people amid celebration, speeches, bands, and fly-by demonstrations nearly twelve years earlier—now sat crippled and hidden, like a wounded animal, somewhere in the Amish Pennsylvania countryside.

Make no little plans; they have no magic to stir mens' blood and probably themselves will not be realized. Make big plans; aim high in hope and work, remembering that a noble, logical diagram once recorded will never die, but long after we are gone will be a living thing, asserting itself with ever-growing insistency.

—"Stirred by Burnham, Democracy Champion"
Chicago Record-Herald, Oct 15, 1910

Epilogue

With those round dips in the wings, it was obvious that Custer, one of aviation's great 'what if' stories, had to be involved. The innovative wing design was his idea, and he alone had pursued it doggedly since the 1940s. He died in 1985, his decades of labor unrewarded.[1]

- Joe Pappalardo

Willard R. Custer died Christmas Day, 1985. With Helen's ever-present help, he had spent the last of his years caring for his wife Lula in their home as she suffered increasing debilitation from Alzheimer's disease. In 1989, Helen buried her by Willard's side at the Rest Haven Cemetery in Hagerstown.

Willard's Channel Wing dream was not impossible, and it was not entirely unrealized. Granted, his dream to revolutionize the aircraft industry and become the Henry Ford of Channel Wing aircraft fell short. But he did succeed in applying his concept of the "Speed of Air" to real world aircraft and came very close to seeing them manufactured and sold.

For example, what if C.G. Taylor had had the money he needed to finish his two-seat Channel Wing prototype? This aircraft design started with the channel and then arranged the rest of the aircraft around it. It would have been small, light, and presenting no "downwash" problem at liftoff. Taylor, a seasoned aircraft designer and builder, had experience obtaining FAA Type Certificates. If he'd had the money he needed, his "flying car" might have revolutionized travel by now, and a daily commute to work from the suburbs might involve dodging drones from the front seat of a Channel Wing.

And Willard's idea lives on, as Robert Englar's Afterword shows. It may yet be included with future air travel.

The only "impossible" aspect of the Channel Wing story is Willard's unyielding perseverance. He started with serious handicaps in his unfinished high school education, his lack of business education or training, and very little money to invest. But he garnered opportunities to learn patent procedures and aeronautical principles from attorney mentors, his son's high school textbooks, and aeronautical engineers, and he accepted advice from successful businessmen. He faced without flinching immense obstacles in lawsuits, skeptical and prejudiced aeronautical engineers and journalists, meddling and thieving investors, and an increasing load of FAA requirements. And then he suffered setbacks in discouraging government studies, serious legal woes, interminable testing costs, and simple bad luck. Yet, he not only persevered, but he inspired his company's officers to persevere until he had spent every effort in his life to realize his dream. This is uncommon and not easily fathomed.

He is a prime example of Teddy Roosevelt's "Man in the Arena." This is the person "who errs, who comes short again and again, because there is no effort without error and shortcoming; but who does actually strive to do the deeds; who knows … the great devotions; who spends himself in a worthy cause; who at the best knows in the end the triumph of high achievement, and who at the worst, if he fails, at least fails while daring greatly."

Willard experienced severe disappointments and made his share of mistakes in the four decades described in this book. Consequently, he did not achieve all that he set out to do. But he did achieve something significant. A banjo-picking car mechanic, he took his homegrown idea and turned it into four full-size aircraft that implemented a radically different approach to flight. Those aircraft were real, and they proved his theories. And in the end, Willard displayed them proudly. They remain his trophies.

Following is a chronological review of each of those aircraft and where they are today, in 2022.

CCW-1

FAA Registration Number N30090

Figure 8.1 — CCW-1 ("Bumblebee") in situ at Silver Hill

Willard built this aircraft, nicknamed "the Bumblebee," in 1941-1942 to specifications provided by Dr. Louis Crook, the Dean of the Department of Aeronautical Engineering at The Catholic University in Washington, D.C. Willard and Crook wanted to see if a full-size airplane with channel wings could fly, and if so, to discover its flying characteristics. Materials (spruce frame and mahogany plywood) and twin Lycoming engines were funded by Briggs Manufacturing Company in Detroit. Willard fashioned the body, channels, and propellers entirely by hand in his backyard shop. Weighing over 1700 pounds, this aircraft flew over 100 hours in tests at the airfield in Beltsville, Maryland where General Gilmore (Army Air Force) saw it and recommended its innovations for wind tunnel testing at Wright Field. Due largely to these tests' results and the testimony of the engineer, Don Young, Willard obtained several patents necessary to continue development. He retired the Bumblebee from service and dismantled it in November 1943. In

1961, he donated it to the Smithsonian Institution, which reassembled it and placed it in long-term storage at the Garber Facility in Suitland, Maryland (Figure 8.1).[2]

As of 2015, the CCW-1 still looked in good shape. As this author learned in private conversation with the current curator, the intent is to move the Bumblebee to climate-controlled storage at the Udvar-Hazy campus of the Smithsonian Institution's Air and Space Museum adjacent to Dulles International Airport in Chantilly, Virginia. However, there is currently no estimated timeframe for that move. Even if moved, it would not likely be on public display.

CCW-2

FAA Registration Number N1375V

Figure 8.2 — CCW-2 Experiment, front view

In 1948, Willard built the CCW-2—a stripped-down Taylor Cub fuselage integrating handmade channels and propellers—in his backyard (Figure 8.2).[3] Like the CCW-1, he intended it solely for experimentation. Based on recommended changes from Don Young, the engineer from Wright Field, it weighed less than the CCW-1 (under 1,000 pounds) and had shallower channels. Flown during 1950 and 1951, the CCW-2 gained notoriety in December 1951 when it rose vertically from the ground while tethered to a windsock pole. In 1952, Willard loaned it to NACA for a wind tunnel study, and then retired it. He sold it for $5,000 in 1954 to an unnamed buyer, and its fate thereafter remains a mystery.

No aircraft were named CCW-3 or CCW-4. The CCW-5 was so-named because it sat five persons.

CCW-5 Prototype

FAA Registration Number N6257C

Figure 8.3 — CCW-5 ("Five-Seven-Charlie") in flight

This aircraft (Figure 8.3)[4] easily qualifies as the workhorse of Willard's effort. Unlike its predecessors, the CCW-5 represented a commercially viable application of channel wings. Baumann Aircraft Company built "Five-Seven-Charlie" in Oxnard, California on a contract funded by John Heinz, II, of Pittsburgh, utilizing a Brigadier fuselage integrating channel wings. Completed in 1953, it became Willard's prototype for his ensuing FAA Type Certification efforts. The most-photographed Channel Wing, this aircraft starred in dozens of demonstrations to military and civilian audiences from California to Virginia, and from Mexico to Newfoundland and Labrador, Canada. All-metal, it was designed to seat the pilot and four commercial executive passengers. The prototype gained instant notoriety when it "hovered" at eleven miles per hour over the runway in a public demonstration in Oxnard on August 24, 1954. In 1967, its reported airtime totaled 241 hours and six minutes. Willard canceled its FAA registration in September 1970, reporting that it was "destroyed," which may be understood as "dismantled."[5] It had not flown at all the preceding year.

Today, a 1/40 scale desktop model of Five-Seven-Charlie, donated by Willard, is on display at the Udvar-Hazy campus of the Smithsonian's Air and Space Museum. It is on the lower level in a glass case housing models of several other aircraft (Figure 8.4).[6]

Figure 8.4 — CCW-5 Model in Udvar-Hazy display case

CCW-5 Production Model

FAA Registration Number N5855V

Figure 8.5 — CCW-5 Modified Production Model (1970)

The CCW-5 Production Model represents Willard's intention to begin production of CCW-5 aircraft based on the CCW-5 prototype, Five-Seven-Charlie, despite lacking an FAA Type Certificate on that model. Furloughed Fairchild Industries employees built it in the hangar at the Hagerstown Municipal Airport using the same blueprints as those used for the prototype. Willard introduced it to the public at "Custer Channel Wing Day," July 4, 1964.

The aircraft served as the basis for renewed FAA Certification efforts under DeVore Aviation. DeVore painted it, then raised its horizontal stabilizers (tail wings) to comply with FAA requirements. It was reintroduced to the public at the airfield in Teterboro, New Jersey on March 16, 1970, after which it was relocated to Enid, Oklahoma for continued study, marketing, and demonstrations. As of July 1970, total flying time recorded to the FAA was 83 hours. By the end of 1971, all demonstration efforts concluded, and the Custer Channel Wing Corporation folded. The aircraft was kept in storage but not maintained until 1975 when Bob Whitehead, president of W.R. Custer Channeled Aircraft, Inc., personally took possession of it and flew it to the Hershey, Pennsylvania area. There it was hidden until after Willard's death in order to protect its design and operational secrets from being stolen.

A year after Willard's death, according to FAA records, the Custer Channel Wing Corporation sold the aircraft to Bob Whitehead. The following year, Bob sold the production model (Figure 8.5)[7] to the Mid-Atlantic Air Museum in Reading, Pennsylvania where it remains today on display outside, subject to the elements (Figure 8.6).[8]

> From the museum, July 2, 2013: "Today the CCW-5 is void of its beautiful base white over polished aluminum with insignia red and navy blue stripes...The Museum maintains the CCW-5 on as-received outdoor display due to space limitations. We hope to begin full restoration in the not-too-distant future."[9]

Figure 8.6 — CCW-5 Production Model (2015)

If you're enjoying this book, why not recommend it to a friend or post a review?

Afterword

Where Do We Go From Here?
The Future of the Channel Wing
by

Robert J. Englar

The unique Powered Lift Aircraft originated and developed by Willard R. Custer as the Custer Channel Wing (CCW) was successfully flown in four different Federal Aviation Administration (FAA) registered aircraft configurations. These aircraft confirmed the novel technical capabilities of this concept as a Vertical Takeoff and Landing (VTOL) or Short Takeoff and Landing (STOL) aircraft; and they were successfully demonstrated to a number of interested possible commercial and military/government partners or customers. However, even though revealed and protected by 26 of his patents, Willard Custer's dream of a commercially successful Custer Channel Wing aircraft was never fully realized in his lifetime or immediately thereafter. Was that because of all the commercial and military interactions and business complexities? Or was it because he hadn't found sufficient sponsors and funds to undertake a full development and FAA-certified program that could lead to a certified, manufacturable, saleable, operational, and maintainable commercial CCW aircraft? In addition to the financial and sponsor issues previously discussed, several possible technical and operation problems arose during Custer's lifetime. Were these solvable and correctable, given an appropriate amount of funding and sponsorship, or were they actually insurmountable? We shall begin to address those issues and their possible solutions in this chapter.

As a much later follow-on to Custer's developments conducted between the concept initiation in 1927 and his final cancellation of the CCW-5 Production aircraft's FAA registration in 1970, there has more lately been much interest in advanced versions of the Channel Wing at such agencies as NASA Langley Research Center (LaRC). Recall that NASA engineers (then called NACA) had previously tested an early version of Custer's CCW-2 in their full-scale subsonic wind tunnel for STOL investigations but had not heavily endorsed it. In the late 1990's, NASA LaRC had become involved with new aerodynamic technologies that could provide performance improvements to current and future aircraft and was aware of certain pneumatic technologies that had recently been or were currently being investigated for or by them. Dr. Dennis Bushnell, then Technical Director of NASA LaRC, showed renewed interest in the Channel Wing [Ref. 1], saying, "The beauty of the Custer Channel Wing is that we can generate lift at zero forward speed by using the engines to provide airflow" over the lifting channels. This of course echoed Custer's early theory. In 2008, LaRC supported this author to pursue and patent further development of advanced Channel Wing technologies and configurations based on these new blowing aeronautical concepts and data developed in the late 1990s and early 2000s [Ref. 2,3,4].

Possible Channel Wing Disadvantages and Problem Areas

The Custer Channel Wing aircraft had confirmed its new VTOL/STOL powered-lift technology3 as lift coefficients of nearly 5 were measured in flight tests; these were greater than 3 times the maximum values for conventional prop-driven aircraft (see Figure F.12 in the Foreword). However, the flight-tested CCW aircraft demonstrated—or perhaps, implied—a number of possible issues or problem areas associated with low-speed handling, cruise drag, stability and control, high-incidence operation, and one-engine-out scenarios. These issues may have been the results of actual low-speed flight tests of this or of other aircraft, of wind tunnel evaluators' data/opinions, or of general-aviation operators' and users' experiences. Some, even most, of these issues may actually be valid engineering evaluations from either wind tunnel or flight tests. These possible drawbacks or shortcomings (we'll call them "Issues" below) could have included:

Issue 1	Much of the CCW's high C_L generated was only from redirected thrust, with less resulting from aerodynamic lift augmentation.
Issue 2	Higher cruise drag could result from additional wing-surface area in the channels.

Issue 3	Asymmetric thrust between left/right engines would yield asymmetric flight moments & instability.
Issue 4	CCW channel leading-edge and trailing-edge flow separation could occur at high angle of attack, α.
Issue 5	Conventional aerodynamic control surfaces showed a lack of low-speed control at STOL speeds.
Issue 6	Nose-down pitch resulted from propeller suction loading on the aft channel segments.
Issue 7	Non-uniform flowfields resulted around the CCW prop at high α with possible vibration and instability.
Issue 8	Poor cruise lift-to-drag ratio and cruise efficiency due to excess channel drag could occur.
Issue 9	High-angle-of attack operation of CCW could cause poor pilot visibility and control.
Issue 10	One-engine-out control problems could occur at low speeds.
Issue 11	Concerns/uncertainties could be formed over new or novel technologies not yet fully integrated into production aircraft.

Possible Channel Wing Improvements

To alleviate these possible "Issues" of a Channel Wing aircraft, preliminary research has been conducted at the Georgia Tech Research Institute (GTRI) under direction of NASA LaRC to adapt Circulation Control (CC) pneumatic blown technology (Figure A.1 and Ref. 2,3,4, for example) to dramatically improve the Channel Wing configuration in any of those Issue areas listed above, if realized.

CC technology has been under development and application for a number of years by the present author to augment aerodynamic lift on both fixed- and rotary-winged aircraft [Ref. 5,6]. CC employs the *Coanda* Effect (Figure A.1) by using small tangential jet blowing over the round or curved trailing edge of the wing to eliminate the aft *boundary layer* (flow velocity shortage) near the wing surface and associated viscous drag. This can greatly augment the circulation *flowfield* around the wing for lift enhancement (CC).

The Coanda Effect involved here was discovered and patented by Romanian inventor Henri Coanda in the early 1930's. It states that a thin jet sheet of fluid emerging onto an adjacent curved surface will follow that surface, entrain (pull in) the surrounding fluid, and create an area of lower pressure on that surface. For a current CC airfoil configuration seen in

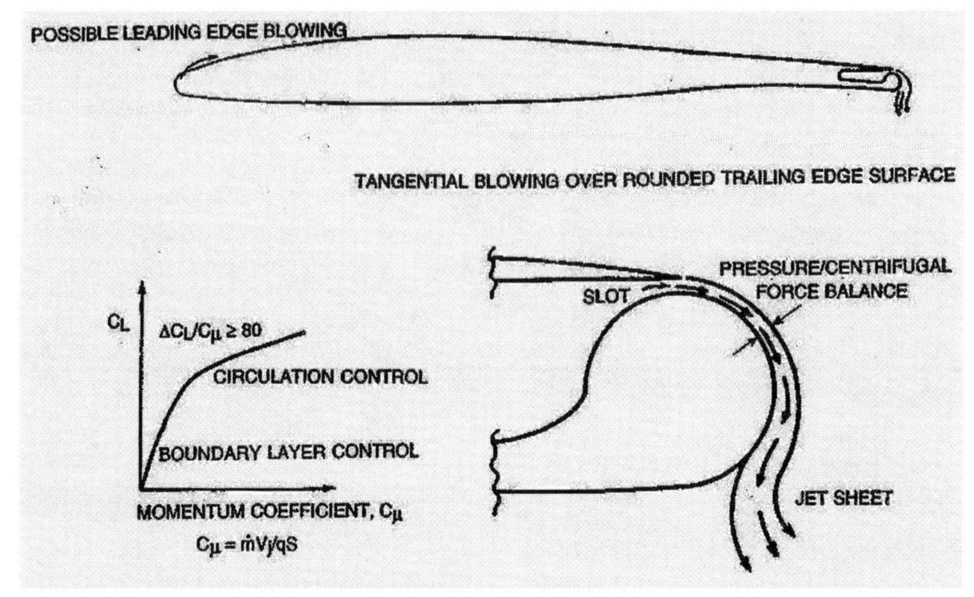

Figure A.1— The Basics of Circulation Control (CC) Pneumatic Technology

Figure A.1, this blowing jet can remain attached (adhered to) the rounded trailing edge until the wing lower surface is reached. This causes the Circulation (a term for mathematical integration of the velocity and static pressure fields around an airfoil to derive the airfoil's lift) to increase, and thus the name Circulation Control (CC). Note in Figure A.1 the balance between the low sub-ambient static pressure on the curved trailing edge surface and the opposing centrifugal force trying to separate the turning jet from that curved surface. The further that this jet can be made to turn around the trailing edge and entrain (pull in) the surrounding flow field, the more it increases the circulation and enhances the lift acting on the blown airfoil (because theoretically, lift = fluid density x velocity x Circulation). This can usually be done independently of the airfoil's angle of attack, so a whole new mode of operational flying is possible. Blowing on this CC airfoil has been shown experimentally to augment the lift coefficient by as much as 80 times the input jet momentum coefficient, Cμ(a non-dimensional aeronautical coefficient defined in the figure). That represents an 8000% return on the invested jet momentum (essentially jet thrust, which equals blowing jet mass flow x jet velocity). This is a huge return on the input blowing airflow, which typically can be bled from existing bleed ports on an aircraft's jet engine. (Note: ΔC_L / $\Delta C\mu$ of only 1.0 would represent a full 100% recovery of the input thrust as a lift increment.) This creation of low static pressures (suction) on the wing upper surface without use of a mechanical flap is very similar to the suction field created by the propeller in the Custer Channel

Wing. This Circulation Control technology has performed exactly as the sketch of lift vs blowing (C_L vs Cμ in this figure (A.1) shows. This is also confirmed by the photos seen in Figure A.2 of the flow visualization tufts on the wing trailing edge of an A-6/CC 1/8-scale wind-tunnel model and the full-scale CC flight test vehicle.

Figure A.2 — (left) Tufts showing CC blowing and jet turning on the 1/8-scale A-6/CC model, and (right) on the full-scale A-6/CC flight demonstrator aircraft (the blowing slot is just at the top of the circular trailing edge)

Results of flight testing on the above A-6/CC demonstrator aircraft [Ref 7,8], using less than half of the readily available bleed air from the stock engines, confirmed maximum C_L values 120% greater than those produced by the conventional A-6's Fowler slotted mechanical flap, or even more applicable, 140% increase in the usable lift coefficient at takeoff/approach angles of attack (α~10°). Also confirmed were 30-35% reductions in the takeoff and approach speeds, resulting in 60-65% reductions in takeoff and landing ground roll distances, and yielding values as short as 600 - 700 ft. This implies that an operational A-6/CC could have flown onto and off of a 1000-foot long Navy carrier deck without the use of arresting cables or a launch catapult. This full-scale confirmation of CC also implied that there was sufficient extra C_L generated to increase the liftable payload by 75% if the conventional takeoff ground roll distance were used instead of STOL. Also shown was that the additional lift-induced drag resulted in much steeper equilibrium glide slopes on approach, where higher engine power settings (which could also be used for quicker response during wave-off) could be offset by this excess drag. In addition to the STOL capabilities confirmed above, an obvious additional improvement is the simplicity of the high-lift, approach-drag-generation, and control-surface systems.

To continue the development of both this proven CC STOL technology and the Custer Channel Wing technology, it seemed a natural for the GTRI and NASA team to combine the two technologies into the Pneumatic Chan-

Upper surface blowing (USB) powered-lift systems—YC-14 mechanical double-slotted flap at left; CCW powered-lift concept at right.

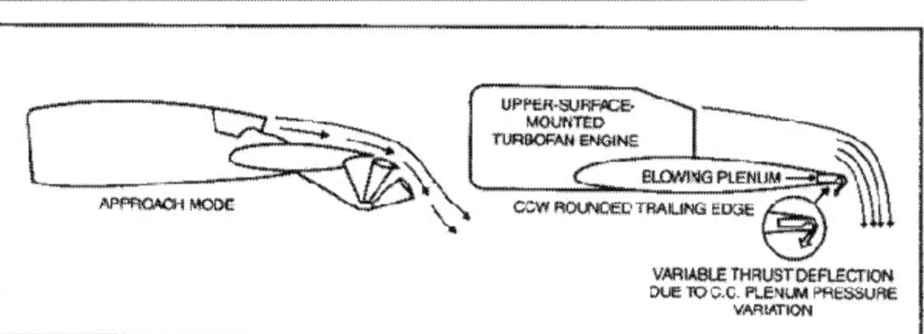

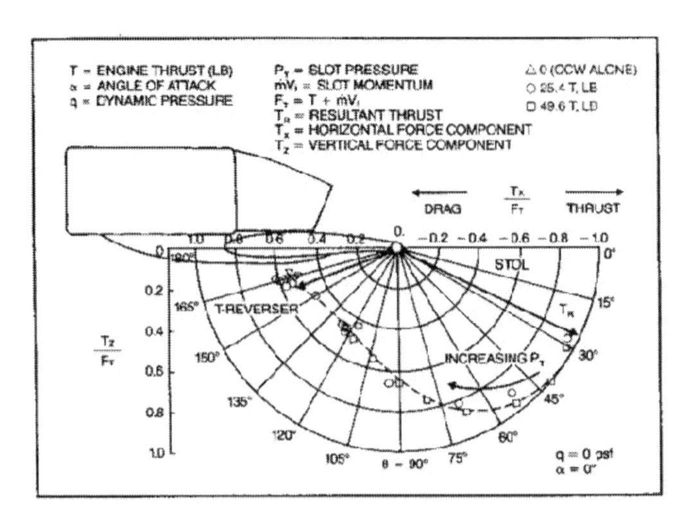

Static thrust deflection of CCW/USB wind tunnel model.

Figure A.3 — (top) Previously Developed Circulation Control Wing/ Upper Surface Blowing Powered-Lift Concept [Ref. 10], and (bottom) model test data

nel Wing (PCW). In this combination (developed and patented by Englar and Bushnell [Ref. 9,10,11,12], an existing powered-lift concept known as CC/ Upper Surface Blowing (USB) was employed. It used jet engines on the upper surface of the inboard wing section (portion closest to the fuselage) and CC on the outboard wing section; this will be the starting point of the new Pneumatic Channel Wing configuration, to be further discussed below.

As Figure A.3 (top) shows, this CC/USB pneumatic configuration combines blowing on curved surfaces at the wing trailing edge (thick arrows) to entrain (pull in) engine thrust to greatly augment the lift and the thrust deflection, even without using high angle of attack. This downward angled engine thrust (thrust deflection) enhances lift in two ways: (1) a vertical thrust force component which equals Thrust x sin (thrust deflection angle + aircraft angle of attack); and (2) flowfield turning that increases the circulation lift on the wing. It also employs blown Circulation Control wing technology on the outboard wing panels (thin arrows) to further augment lift and low-speed controllability while providing **additional drag** when needed for slow-speed approaches down steep glide slopes for Super-STOL, or while **reducing drag** for efficient climbout and cruise.

This engine thrust turning/deflection and lift augmentation are based on the Circulation Control/Upper Surface Blowing (CC/USB) concept depicted in Figure A.3, but the earlier original USB concept was developed and flight tested by NASA on its Quiet Shorthaul Research Aircraft (QSRA) [Ref. 10,11] as a powered-lift demonstrator aircraft system. The QSRA employed multi-element mechanical flaps to entrain/deflect engine thrust (Figure A.4). NASA had developed this QSRA STOL flight demonstrator on a De Haviland Buffalo C-8A turboprop aircraft. In this USB configuration, the 4 turbofan engines are located on the wing's aft upper surface, with D-shaped nozzles spreading the engine exhaust over the flaps. Again, the suction over the flaps entrains the engine exhaust and directs it downward. Powered lift thus occurs because of the higher exhaust velocity over the flaps (i.e.,

Figure A.4 — (left) NASA QSRA Aircraft, Aft View (Johan Visschedijk Collection photo); (right) Front View, STOL (NASA photo)

Custer's "speed of air") aided by the deflected thrust's vertical component. Using mechanical flaps, this NASA proof-of-concept aircraft flight-demonstrated extreme STOL capability on short runways and was even able to takeoff and land on a Navy aircraft carrier without use of either the launch catapult or the arresting cables—amazing for a transport of that size. Ground-based STOL was also later demonstrated by the prototype Boeing YC-14, which also employed USB on its mechanical flaps.

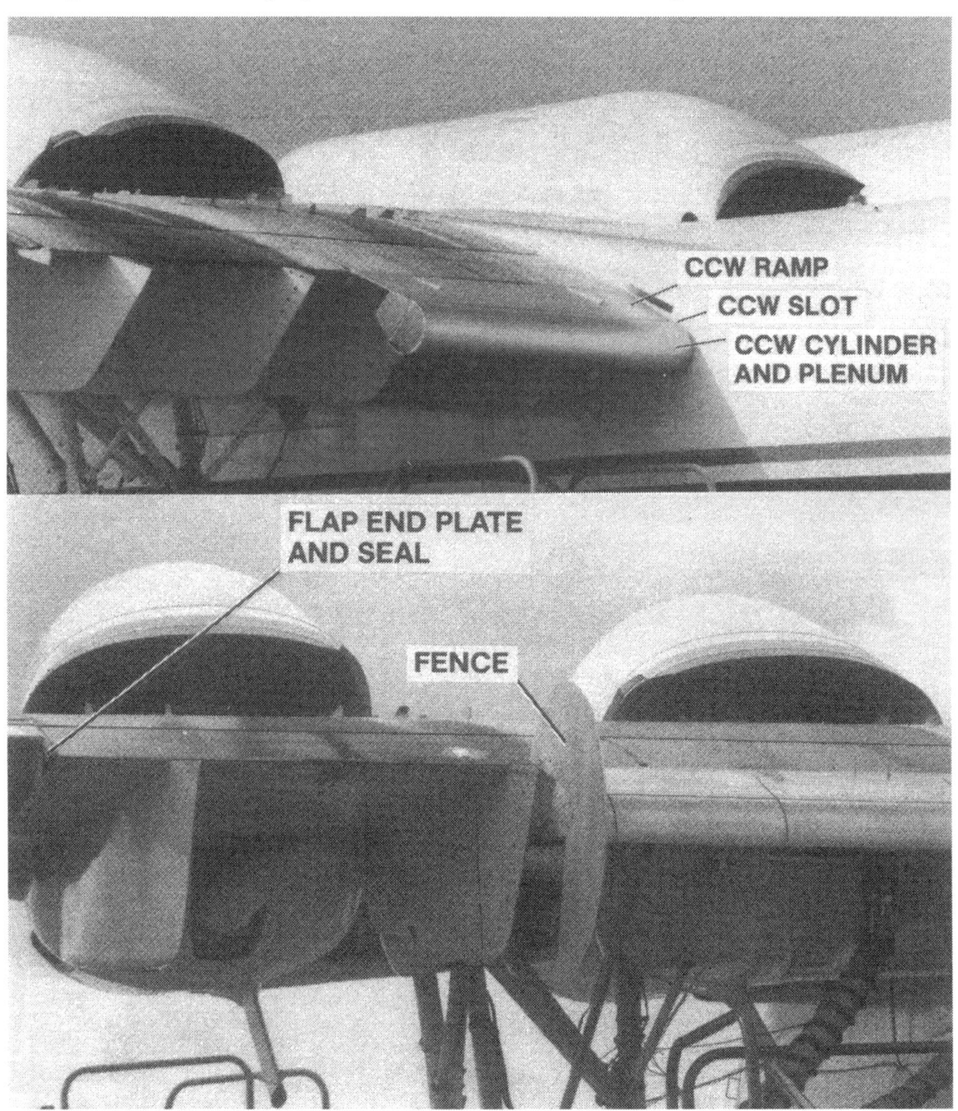

Figure A.5 — CC/USB trailing edge installed on QSRA inboard wing (only) for a static thrust turning test at NASA Ames Research Center [Ref. 11,13]

To evaluate the Circulation Control blown aerodynamics combined with USB technology full-scale, a proof of concept CC/USB demonstrator was built and tested statically on one wing of the existing QSRA aircraft (Figure A.5) [Ref. 11,13,14]. It had a CC circular trailing edge and blowing slots that were straight spanwise, with the engine exhaust located above and straight ahead of the slots. The CC trailing edges with different arc lengths were interchangeable during these ground tests. Thrust deflection angles as high as 165° produced by blowing had already been measured experimentally on wind-tunnel models (Figure A.3) [Ref. 10,11]. These model CC/USB concept tests had provided pneumatic vertical thrust deflection of 90° for VTOL at intermediate blowing total pressures (PT), and also confirmed thrust-reversing capabilities for STOL, STOVL or VSTOL at higher blowing levels, without any moving parts. Remember, the CC wing alone (Figure A.1, A.2) had employed a similar tangential-blowing configuration, but without the pneumatic thrust deflection. Also remember, these blown CC wing airfoils alone have generated measured 2-D lift augmentations of 80 times the input blowing momentum [Ref. 5,6,7]. In addition, the CC lift and drag capability outboard can be applied differentially to generate very large rolling and yawing moments, which are essential for controlled flight at the very low speeds of Super-STOL. These CC/USB ground tests employed both a 180° round trailing edge CC and a 90° arced CC trailing edge (see Figure A.5) on the NASA QSRA aircraft, and verified thrust deflection angles of over 100° [Ref. 13,14]. These deflection angles were measured statically on a ground balance stand and thus included no aerodynamic lift components (no wind or flight), but they did verify the model thrust-turning test results of Figure A.3.

Pneumatic Channel Wing (PCW)

From these results, it became obvious to Englar and Bushnell in the Ref. 9 patent (in 2006) that the Custer Channel Wing concept could be integrated with the Circulation Control wing and Upper Surface Blowing (CC/USB) by replacing the upper surface USB engines with the prop-driven channel wing to become the new Pneumatic Channel Wing (PCW) configuration.

Based on the earlier CC/USB wind-tunnel and full-scale data (Figure A.3) [Ref. 10,11,13,14] and CC wing alone flight test data from the A-6 STOL-demonstrator program [Ref. 8], the predicted lift and drag capabilities for the proposed Pneumatic Channel Wing configuration were expected to offer great Super-STOL promise. Reference 12 details these early predictions before the later wind-tunnel test data were available. These suggested C_L values approaching 9-10 for a Pneumatic Channel Wing aircraft with blowing on outboard CC wing panels at relatively low aircraft angle of attack. Due to the additional vectored thrust component, higher lift values were possible with higher thrust levels. Again, for comparison, the Custer Channel Wing aircraft generated an in-flight C_L of 4.9, while a conventional slotted flap on this wing geometry could generate C_L from 2 to 3. (Should a Pneumatic Channel Wing be considered, a turboprop—jet turbine powered—engine might be needed to also provide the pressurized blowing air for CC on the channel or the outboard blown wing. Of course, an auxiliary air supply like an APU could be substituted for this purpose.) The initial STOL takeoff predictions of Ref. 12 showed that these PCW capabilities could produce very-short takeoff ground rolls for typical mission weights even on a hot day when takeoff performance is usually worse (longer runways required), and even yield zero ground roll under certain conditions.

As part of the program for the NASA Langley Research Center to develop this Pneumatic Channel Wing (PCW) concept incorporating a blown channel wing, GTRI and NASA teamed in an experimental development effort. This provided aerodynamic and propulsive data input for design studies conducted at both NASA and GTRI. The following pages summarize these experimental results and discuss results from variations in PCW channel/wing configuration geometry, propeller thrust, and channel blowing. A major goal of this program was to resolve as many as possible of the above listed potential problem areas (Issues 1-11) seen on earlier channel wings by employing a pneumatic channel wing (PCW) powered-lift configuration, and we will discuss these successful resolutions below.

Wind-Tunnel Evaluations and Results for Pneumatic Channel Wing model

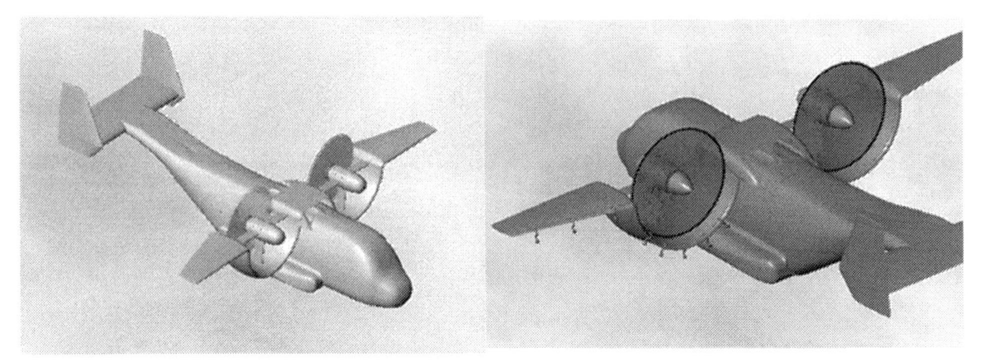

Figure A.6 — Conceptual Pneumatic Channel Wing Super-STOL Transport Configuration

A wind-tunnel development and evaluation program was conducted at GTRI on the generic twin-engine Super-STOL-type PCW transport design depicted in Figure A.6, using the 0.075-scale half-span model shown in Figure A.7. This program was conducted in GTRI's subsonic research wind tunnel.

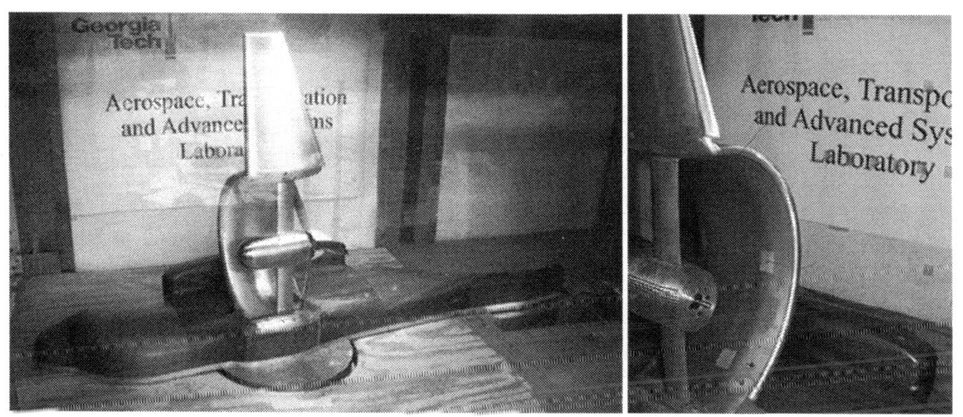

Figure A.7 — Pneumatic Channel Wing Semi-span Model Installation in GTRI Model Test Facility Research Tunnel (3-bladed prop with Unblown Outboard CC), plus Jet Flow Turning in Channel (see black tufts)

Here, a variable–speed electric motor was installed in the nacelle. It drove interchangeable 2-bladed, 3-bladed or 4-bladed propellers of various diameters and pitch. Also variable was the height of the blowing slot located at 95% of the channel chord length, as well as the blowing momentum coefficient and portions of the slot arc length which were blown. Behind the slot, the rounded trailing edge curved only 90° (rather than the more conventional 180° of typical CCW configurations shown in Figure A.1 and A.2) for an anticipated maximum thrust deflection of around (90° + α). We already knew (Figure A.3) that thrust deflections up to 165° yielded by blowing were a possibility. Here, the blowing momentum coefficient is defined as (m is jet mass flow and V_j is jet velocity):

$$C\mu = m V_j / (qS)$$

This semi-span model configuration (Figure A.7) was mounted on an under-floor balance with air supplies and an automated pitch table in the GTRI Model Test Facility's 30" x 43" x 90" wind tunnel test section. The emphasis in the following data is on the performance of the inboard blown Pneumatic Channel Wing configuration, but performance of the outboard CC sections to further augment lift will also be shown.

Some 980 wind-tunnel runs were conducted during three test programs at GTRI to develop these blown-configuration aerodynamic details/geometries and to evaluate their aero-propulsive (wing/propulsion interactions), flight-trim, and control characteristics. Typical test results are presented in the following sections to demonstrate how the blowing and thrust variations affected overall performance.

Tunnel Test Results, Outboard CC Wing Installed

Adding the propeller in the channel (Figure A.7, left photo) brings into play the powered-lift characteristics of the Pneumatic Channel Wing configuration. The right photo shows earlier testing of the aerodynamics of the blown channel wing without the prop. The black tufts at the channel confirm 90° of jet turning there due to blowing. Figure A.8, for $\alpha = 0°$ angle of attack, shows the variations in lift and drag with thrust variation for fixed values of blowing coefficient. Here, in order to recognize the direct thrust components to lift and drag, thrust coefficient is defined as $C_T = T/(qS)$, where T is the calibrated uninstalled wind-on prop-alone (not-in-the-channel) thrust. The reference area S is the wing half-planform area, including the channel. More details of test technique are found in Ref. 10.

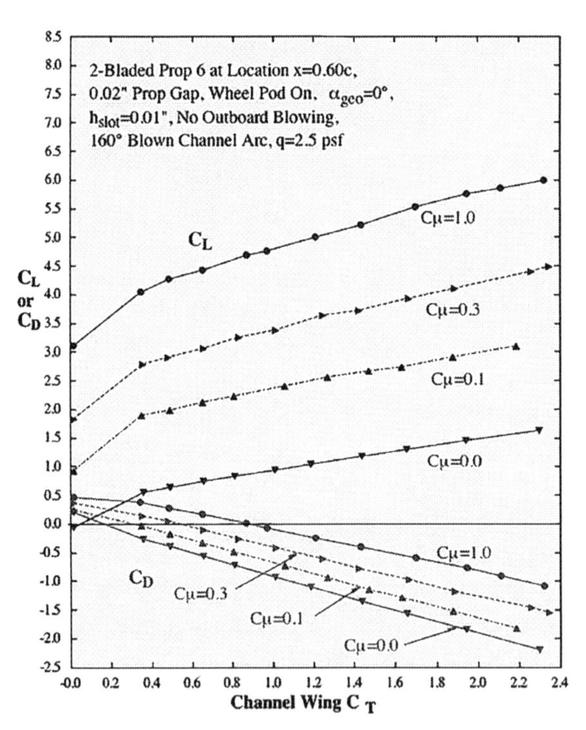

Figure A.8 — Effects of Prop Thrust Variation on Lift and Drag at Constant Blowing (Cμ) and α = 0°, CC wing installed but not blown

In this Figure, the aerodynamic coefficients C_T, C_L and C_D are directly comparable so that one can determine forces resulting from channel wing thrust. Here at 0° angle of attack, lift coefficients of double to triple the C_{Lmax} of conventional flaps are generated by thrust and blowing, independent of angle of attack change. Also, note that measured drag coefficient C_D includes the input thrust, which cannot reasonably be separated from the aerodynamic drag alone once the prop is in the channel. Measured drag can thus be (and sometimes is) negative, headed forward. This is perfect for STOL takeoff. After the initial low values of thrust are exceeded, C_L increases nearly linearly with thrust, and C_D reduces nearly linearly, which is great for control systems. (This implies that at constant blowing, the thrust deflection angle is nearly constant.)

Figure A.9 shows that for $\alpha = 10°$, incremental lift augmentation due to blowing ($C\mu$) is much greater than due to thrust. Here at $C_T = 2.2$, the blown configuration generates C_L around 8.5 at $\alpha = 10°$. The flight-tested Custer Channel Wing [Ref. 15,16] generated roughly 1/3 this value of C_L at this thrust level, but also required an angle of attack of 24°-25°. Note also that increased blowing at constant thrust yields increased drag (rather than thrust recovery), which can be quite useful for Super-STOL approaches and short landings. These lift comparisons in Figures A.8 and A.9 show that lift increases more efficiently by increasing blowing than by increasing thrust.

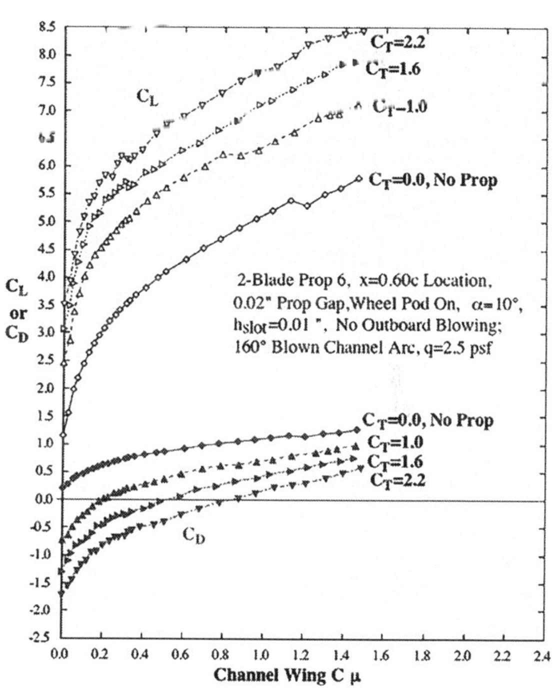

Figure A.9 — Effects of Blowing Variation on Lift and Drag at Constant Thrust (C_T) and $\alpha = 10°$

The Figure A.10 plot shows the variation in lift and drag with angle of attack α for the blown powered-lift configuration in comparison to the unblown baseline configuration without the prop. Here, flow visualization showed that the initial stall (α = 15°-17°) seen for most of the lift curves corresponded to stall of the outboard unblown wing section, while the blown channel wing section then continued on to stall angles of 40°-45° and C_L values of 8.5 to 9. Notice that C_D (including thrust) increases from negative to positive values as angle of attack is increased.

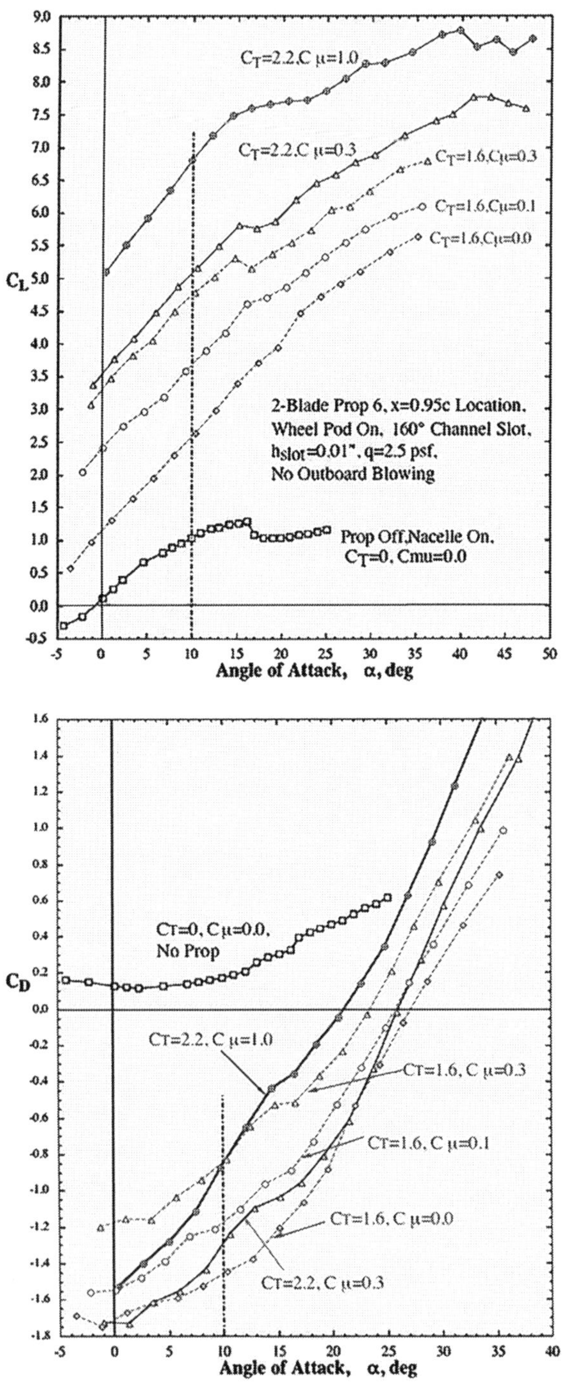

Figure A.10 — Effects of Blowing, C_T, and α on Lift Coefficient, Stall Angle and Drag Coefficient for the Pneumatic Channel Wing Model with Unblown Outboard Wing

Tunnel Test Results, Channel Wing Only (outboard CC wing removed)

Higher non-dimensional thrust coefficient values were desired for testing and were provided by the test model when the channel-only configuration was tested (fuselage, blown channel and prop, but with no outboard CC panels), because then the reference area of the wing (in the coefficient denominators) was reduced to only that of the channel wing section. This allowed C_T of ~3 for the channel-only vehicle, and as Figure A.11 shows, lift coefficients nearing 11 were measured. **This of course is more than double the powered lift of Custer's CCW-5, even though no outboard wing is present here!** Needless to say, not all of the lift values shown in Figure A-11 are trimmed (i.e., pitching moment = 0.0).

The possible inability to trim out the pitch of these Super STOL aircraft was pointed out as a problem of blown systems in Ref. 9,10, and in the channel wing problem areas (Issues) listed above. It is further emphasized in Figure A.11 test results, because the large suction on the aft loaded blown channel (as well as blown wing) produced large nose-down pitching moments. Even though this can produce improved longitudinal flying stability, these moments must be trimmed (to yield no pitching moments). Horizontal tail investigations were conducted (see Refs. 2,3,10) as part of this 3-D model development in hopes of determining tail location and configuration to provide enough nose-up pitch to trim the vehicle. Testing of tail-on configurations revealed that a low-tail position immersed in the prop slipstream was the most effective for pitch trim, which was similar to Custer's tail control discoveries. However, at very high thrust, high blowing and high lift, there were indi-

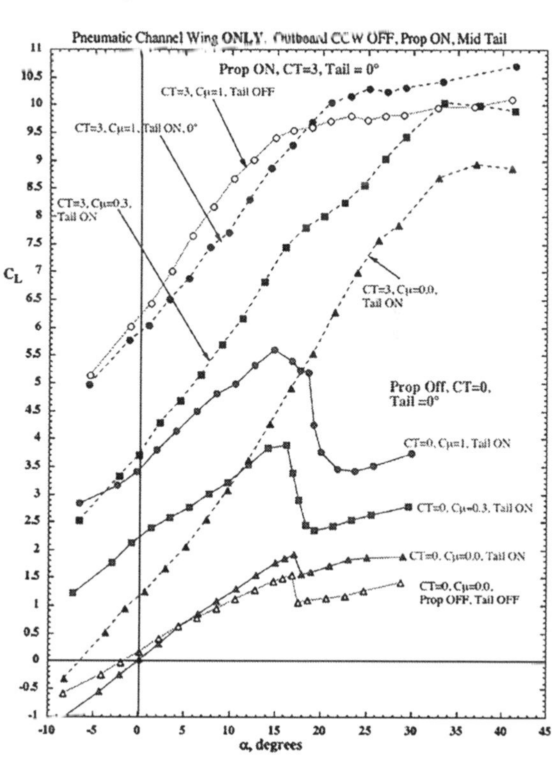

Figure A.11 — Effect of Thrust and/or Blowing Increase on Lift Variation with α for Channel-Wing-Only Configuration (No Outboard Wing Panels)

cations that tail stall could still be a problem. Thus, this data implies that further tail development (perhaps including tail leading-edge blowing to prevent the tail stall without mechanical LE fixes) is needed to trim in this high C_L range.

Tunnel Test Results: Flow Attachment (Methods to keep Flow Fields from Separating from the Wing or Channel)

An additional series of flow visualizations was conducted to further identify means to prevent separated flow fields (or stalls) on the aircraft during high-lift generation. Figure A.12 data show that the flow at the channel leading edge is entrained by the prop to the point where LE separation is prevented up until α = 35°-40° or more, but that the outboard CCWing is prone to stall there. Leading-edge (LE) blowing on this outboard CCWing panel was found to greatly entrain this flowfield as well, with stall not occurring until above 50°. However, this high-alpha mode of operations is not the intended mode to obtain high lift, merely a means to prevent unexpected flow separation and stall in a sudden unexpected upwash/updraft. Flow visualization in Figure A.13 shows this severe separation on the outboard wing at α = 20° for the unblown LE case (left photo), while blowing the wing leading edge completely re-attached the flowfield there. The channel flow itself remains attached **without any blowing!!**

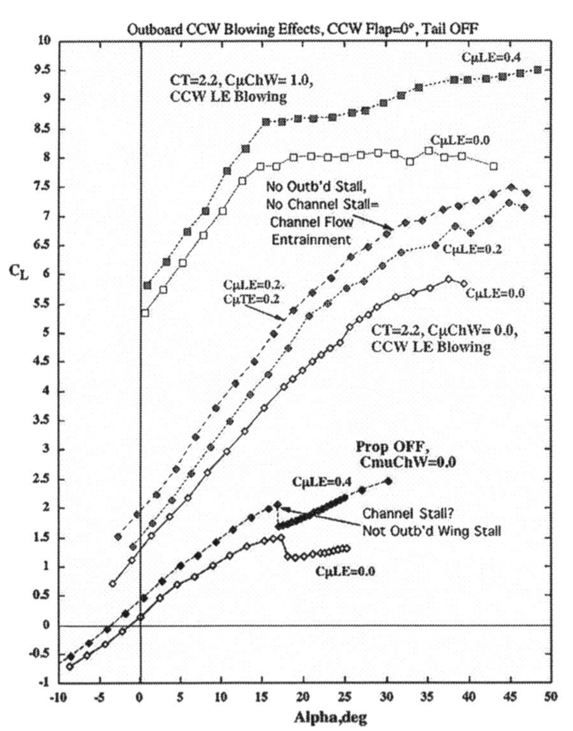

Figure A.12 — Leading-edge Blowing and Channel Flow Entrainment Prevent Flow Separation over Both Channel and Outboard CC wing Leading Edges

An additional means of trim and control was investigated for the Pneumatic Channel Wing. This involves offsetting these large nose-down

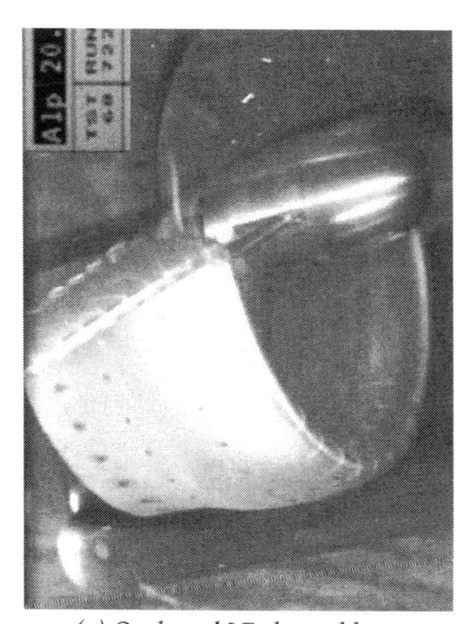

(a) Outboard LE slot unblown

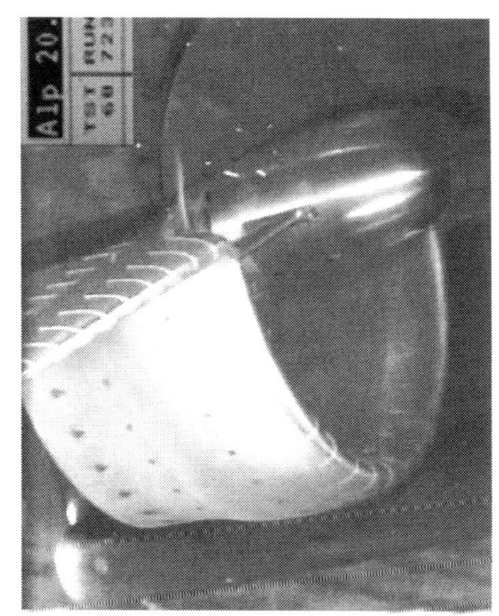

(a) Outboard LE slot blown

Figure A.13 — Flow Attachment caused by leading-edge blowing on outboard CC wing and channel flow entrainment at $\alpha = 20°$. Channel leading edge is not blown in either photo

pitching moments by moving the aircraft center of gravity (cg) further aft to trim. Aft cg movement was previously performed for flight tests of the A-6/CC Wing aircraft (Figure A.7). Some small control surface (such as a blown canard to provide nose-up pitch and associated positive lift to trim) could perhaps be incorporated to make this a feasible pitch-trim device, without experiencing the typical STOL aircraft problem of lift loss due to tail download to trim.

Comparison of Wind Tunnel Measurements and Predictions

In Figure A.14, the test results from the above wind tunnel investigations for the PCW with the outboard CC wing installed (dashed curves with data points) are compared with previously-predicted lift and drag data (solid curves without data) which were **estimated** for this PCW from existing CC/USB wind-tunnel data and from A-6/CCW flight-test data. Whereas the prop/electric motor currently available on this tunnel model did not allow higher experimental C_T values than about 2.2 (outboard wing installed), this lower-thrust wind-tunnel data considerably surpasses the predicted lift data (Figure A.14(a)). If the ratio of measured-to-predicted holds proportionally up to C_T=10, then aircraft C_L values over 14 can be expected for PCW at the low angle of attack of only 10°. For reference, Custer's CCW-5 flight test (see

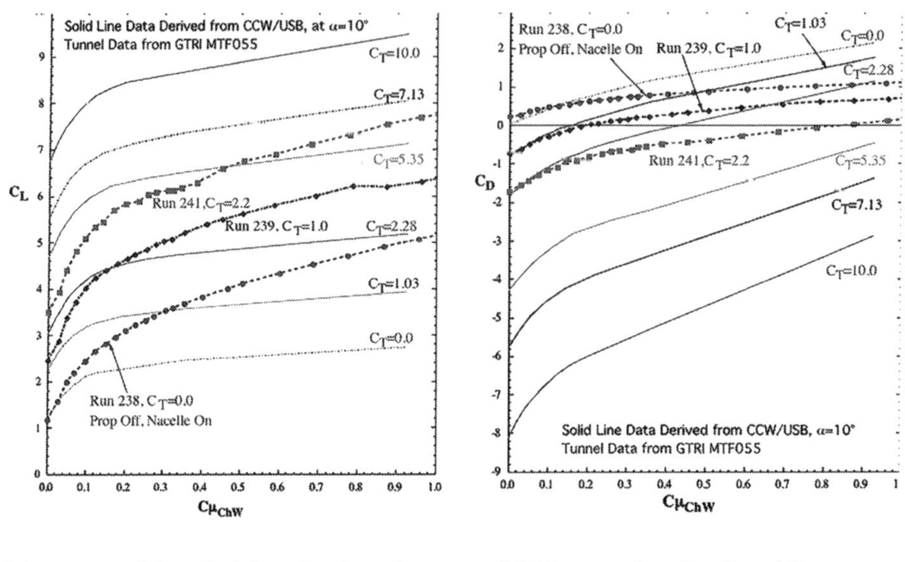

(a) Measured (symbols) vs Predicted Lift (no symbols) *(b) Measured vs Predicted Drag*

Figure A.14 — Comparisons of Predicted and Experimental PCW Lift & Drag Data at Constant C_T, $\alpha = 10°$, Outboard CC ON

Foreword, Figure F.12) had demonstrated Power-on C_L near 5.0 but at much higher angles of attack. The experimental drag data (Figure A.14(b)) are similar to the predicted values at lower blowing but show lower drag than was predicted at higher blowing. The measured-versus-predicted results in Figure 14 seem thus to suggest that even better takeoff performance might be obtained (due to higher lift, lower drag) if measured data rather than estimated aero/propulsive data had been used. However, the lower measured drag values indicate that additional attention will need to be paid to obtain ***greater*** drag values for steeper glide slopes on STOL approaches (when/if desired and chosen by the pilot).

These estimated data were used to predict Super-STOL takeoff distances for a postulated CC/USB aircraft on a 91° hot day at 3000 ft altitude [Ref. 11,12] (Figure A.15 and A.16). These takeoff ground roll data in Figure A.16 compare a nominal conventional tilt rotor aircraft, similar to the V-22 Osprey, to the same size and weight PCW powered-lift aircraft (Figure A.15). Instead of hovering, the tilt rotor is inclined to 51.5° tilt angle for a STOL short takeoff.

The postulated PCW shown in Figure A.15 employs an outboard blown CC wing and two representative wing areas. Both the baseline tilt rotor and the PCW are predicted to achieve a 100-foot takeoff ground roll, but at

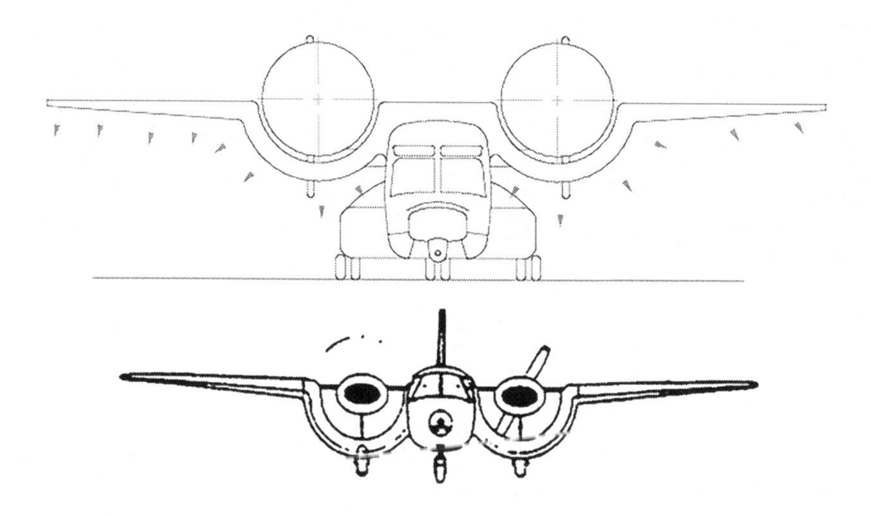

*(a) - Layout of Pneumatic Channel Wing compared to Custer CCW-5 (not to scale):
PCW (top) wing area = 381.35 sq. ft., wingspan = 58.0 ft. prop diameter = 108 in.;
Blowing on Channel Wing and on outboard CC wing at sites indicated.
CCW-5 (bottom) wing area = 209 sq. ft., wingspan = 41.0 ft.*

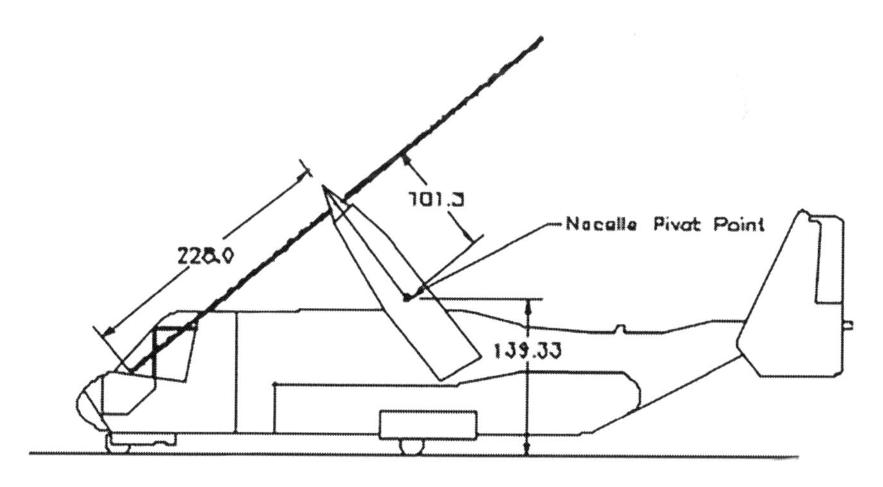

(b) - Generic Tilt Rotor, STOL mode, 51.5° rotor tilt angle, wing area = 381.4 sq. ft.

Figure A.15 – Pneumatic Channel Wing (a) and Generic Tilt Rotor Baseline Aircraft (b)used in STOL Takeoff and Landing Analyses

different gross weights. With a 10% increase in thrust (engine scaling factor of 10% increase in size, ESF = 1.1), the PCW is predicted to be able to achieve the 100-foot STO takeoff with a 17,000-pound increase (41%) in takeoff gross weight.

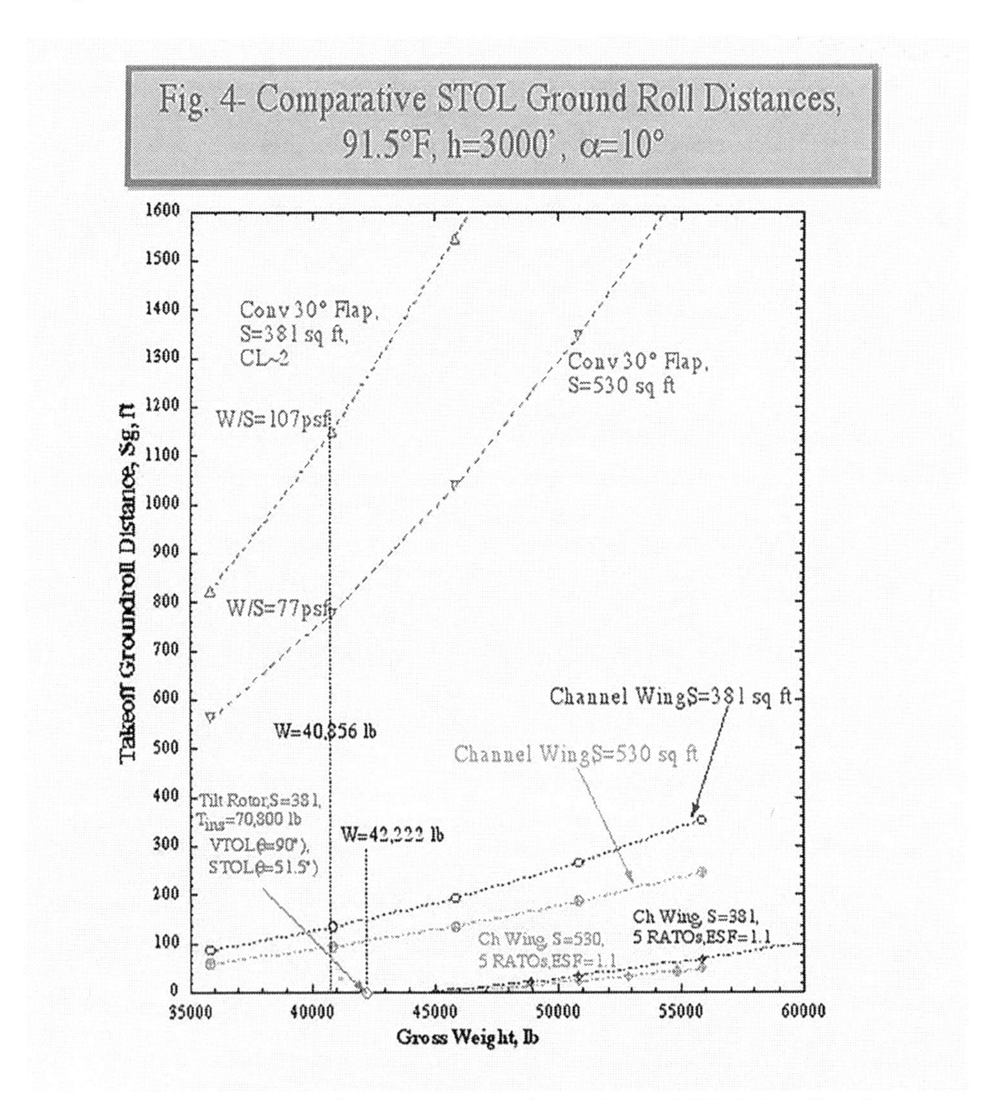

Figure A.16 — Pneumatic Channel Wing Predicted Super-STOL Takeoff Performance NOTE: ~17,000 lb. Increase in TO Gross Wt. over Baseline Tilt Rotor at 100' ground roll

Potential Applications of Future Pneumatic Channel Wings

Design and mission studies conducted at NASA LaRC based on the above tunnel data have led to consideration of several new pneumatic powered-lift PCW-type configurations. The capability of the Pneumatic Channel Wing to significantly augment lift, drag, and stall angle to the levels reported herein demonstrates that this technology could enable simple/reliable/effective STOL and possibly VTOL operations of personal and business-sized aircraft operating from remote or small sites as well as increasingly dense urban environments. Such capability now opens the way for alternate visions regarding civilian travel scenarios, as well as both civilian and military aerial missions. One such vision is represented by Personal Air Vehicle Exploration (PAVE) activity at NASA Langley Research Center. Another vision, a military Super-STOL transport, is discussed in the mission study of Ref 12 and Figure A.16, above. It is suggested that a blown channel wing with props could replace the upper surface blowing inboard wing section of the USB aircraft of Figure A.3. This could augment the powered-lift even more, as well as augment the vertical engine thrust turning to improve the already existing VTOL capability.

Summary and Conclusions relating to future Pneumatic Channel Wing aircraft

Results from these subsonic wind-tunnel investigations conducted at GTRI on a powered-lift 0.075-scale powered semi-span model of a conceptual Pneumatic Channel Wing (PCW) transport have confirmed the potential aerodynamic payoffs of this possible Super-STOL configuration, including very high lift and overload capability. These results/benefits include the following (see Ref 2,3,4,12 for more details). **Also shown with each result/benefit below is notification of the solution(s) to the potential Issues and/or problem areas originally anticipated for the Custer Channel Wing and enumerated early in this chapter:**

A. Blown/powered wind tunnel and flight investigations have confirmed Circulation Control and the Pneumatic Channel Wing (PCW) concepts for **very high lift generation enabling Super STOL** and overload capability using **no moving aerodynamic elements.** Pneumatic/propulsive high lift incorporated with blown thrust deflection have generated C_L=10-11 at low angle of attack to avoid high α operations, flow separation and loss of control. **Drag coefficient varying between negative and positive** values were seen and can aid in either Short Take Off or Short Landings depending on blowing and thrust levels. **C_L's nearing 14 are predicted for the PCW,** if higher thrust is available, for example, on takeoff.

Solutions to:

- **Issue 1** (high lift from other than pure thrust deflection)
- **Issue 2** (high cruise drag due to channel wing)
- **Issue 8** (poor cruise lift-to-drag ratio)

B. Blowing and thrust increases were both found to significantly enhance circulation, thrust deflection and lift; but, if evaluated as incremental lift per unit of input thrust or momentum, **blowing was far more efficient than thrust.**

Solution to:

- **Issue 1** (high lift from other than pure thrust deflection)

C. By varying only blowing and/or thrust, **all the aircraft's aerodynamic characteristics (3 forces and 3 moments) can be augmented or reduced** as desired by the Super-STOL aircraft's pilot or its control system **without mechanical moving parts** (such as tilting rotors or wings) and **without resorting to high α** to acquire larger vertical thrust components for lift.

Solution to:

- **Issue 1** (high lift from other than pure thrust deflection)
- **Issue 5** (conventional aerodynamic control surfaces lack of low-speed control power/authority)
- **Issue 6** (nose-down pitch control)
- **Issue 10** (one-engine-out control problems)
- **Issue 12** (complex mechanical control elements)

D. The blown channel wing itself, without thrust applied, was able to double the CLmax capability of the baseline unblown aircraft configuration and multiply its **lift at $\alpha = 0°$ by a factor of 10.** Addition of blowing on the outboard CC wing section can increase this further and can also add drag as needed for Super-STOL approaches. This implies **low angle of attack STOL** operations.

Solution to:

- **Issue 1** (high lift from other than pure thrust deflection)
- **Issue 2** (high cruise drag due to channel, plus need for high

drag on STOL approach)

- **Issue 4** (leading-edge and trailing-edge flow separations at high angle of attack)
- **Issue 7** (non-uniform flowfield around props at high angle of attack, resulting vibration)
- **Issue 9** (poor pilot visibility and control at high angle of attack operations)

E. Even with the unblown outboard wing stalling at lower angle of attack, the blown and thrusting channel continued to increase lift up to a stall angle of 40°-45° due to channel flow entrainment. While this high α may not prove practical as a takeoff/landing operational incidence, it does show significant **improvement over the asymmetric leading edge separation** of the conventional channel wing's stalled channel and the resulting low-speed control problems.

Solution to:

- **Issue 3** (asymmetric thrust from left/right channels)
- **Issue 4** (leading-edge and trailing-edge flow separations at high angle of attack)

F. PCW conversion of engine thrust into either **drag decrease or drag increase** without moving parts is also quite promising for STOL takeoff or landing operations.

Solution to:

- **Issue 2** (high cruise drag due to channel)
- **Issue 3** (asymmetric thrust from left/right channels)
- **Issue 8** (poor cruise lift-to-drag ratio)

G. Large **nose-down pitching moments** produced by these blown configurations, and thus low-speed flight control longitudinal trim capability, will need to be addressed in future evaluations.

Solution to:

- **Issue 6** (nose-down pitch control) not yet totally resolved for PCW, but differential blowing between fore and aft blowing slots on CC wing can begin to address this issue.

H. Unlike a tilt rotor, in Super-STOL or VSTOL, **there is no download on the wing** from prop thrust since the PCW props don't tilt.

Solution to:

- **Issue 13** (No wing download from propulsor is possible for PCW, since propulsor doesn't tilt over the wing)

I. Pneumatic Channel Wing's potential for an **integrated lift, thrust/drag interchange and control system** all from one STOL system holds promise in terms of simplicity, weight reduction and reliability/maintainability.

Solution to:

- **Issue 11** (novel new technologies not readily accepted by aviation community and users)
- **Issue 12** (complex mechanical control elements and systems)

Thus far, the projected operational benefits based on these early data suggest a Super-STOL and possible VSTOL capability for PCW with significantly increased payload, reduced noise signatures, and increased engine-out control, all without variable geometry or mechanical engine/prop tilting. As seen in the above listing of demonstrated results and benefits, potential solutions to the *entire list of original possible Issues or problem areas can be provided by the capabilities of the Pneumatic Channel Wing configuration.* A Pneumatic Channel Wing aircraft thus equipped could provide a simpler, less costly way of achieving Super-STOL/VSTOL capability without the complexity, weight or reliability issues of existing tilt rotor aircraft, which rotate the propulsion system, carry large engines and rotors on the wing tips, or thrust downwards on their fixed wings during hover. Additionally, the **integration of pulsed-blowing technology** with Circulation Control (more recently investigated [Ref. 17]) **may further increase lift efficiency and reduce already low blowing requirements by up to 50%** or more, while further enhancing stability and control. The remaining issue is Issue 11: Concerns/uncertainties over new or novel technologies not yet fully integrated into production aircraft, which can only be resolved with continued developmental effort for the Pneumatic Channel Wing. This would indeed be a worthwhile follow-on to advance the promises of Willard Custer's original Custer Channel Wing aircraft.

Successful application of these results can lead to positive technology transfer to personal, business, and military sized aircraft. In addition to the military Super-STOL transport discussed above, these experimental data and pneumatic technology results have been included in preliminary design studies of other possible pneumatic powered-lift configurations, including smaller personal and business-type aircraft. **These pneumatic improvements shown by the Pneumatic Channel Wing are directly related to the early efforts of Willard Custer on his Custer Channel Wing, on his proof of this technology in hover and STOL flight, and on his identification of possible issues that must be resolved.**

Recommendations, or "Where Do We Go from Here?"

The above results and conclusion for Pneumatic Channel Wing have continued the much earlier developments by Willard Custer of his Custer Channel Wing technology in the timeframe from 1927 to 1970. We now need to extend them beyond the above more recent Pneumatic Channel Wing configuration. **Additional testing, evaluation and development still need to be accomplished to address and reduce any possible operational issues such as pitch-trim problems, performance at higher thrust levels and lower blowing, associated moment trim and stability and control, and required blowing air supply.** In the future, the existing PCW wind tunnel model or larger 3-D powered-lift models should be modified to include blown tail surfaces or blown forward canards, and additional improvements to the pneumatic thrust deflection powered-lift, control, and air supply systems. The following should be experimentally developed and investigated:

- Use of **pulsed blowing** to further reduce required blowing mass flows (both inboard on the channel wing and outboard on the Circulation Control wing).
- **Greater thrust** and powered lift, or improved propeller characteristics for greater C_T availability.
- Further evaluation of **low-speed controllability** and trim, including evaluation of improved tail surfaces, which might even be blown to reduce tail area and drag.
- Additional evaluation of low-speed controllability and trim by novel **aerodynamic/ pneumatic trim and control devices** (blown canards, for example.)
- Additional evaluation of the Channel Wing propulsion system. Do we need to substitute a turboprop for Custer's internal combustion engine in order to improve fuel consumption and cruise efficiency, as well as to provide a source(s) of compressed air for the blown Pneumatic Channel

Wing and outboard blown Circulation Control wing? Or do we need to remove the conventional propellers entirely and substitute Turbofan engines, much like the CC/USB configuration or the NASA QSRA aircraft?

The earlier STOL mission analyses should be revised to incorporate the experimentally developed aeropropulsive and stability & control characteristics of the Pneumatic Channel Wing concept. If these projected benefits are confirmed, and further benefits come to light, then larger-scale, higher-Reynolds-number testing on a full-3-D Pneumatic Channel Wing model should be conducted to make greater strides toward this pneumatic powered-lift technology's maturation.

The ultimate goal, of course, is to develop, flight test, FAA certify, and then manufacture a commercially successful advanced Custer Channel Wing that Willard Custer had always hoped and worked for, to once again confirm that it's "the speed of the air induced or entrained over the wing/channel that is so important, and not the air speed of the associated aircraft."

Photos

Figure A.4 - Johan Visschedijk, 1000aircraftphotos.com (by permission), and NASA QSRA, NASA Ames Aviation Services Division, Public domain, via Wikimedia Commons, color removed.

References

[1] Wright, Tim, "That Extra Little Lift -- Has Aviation Finally Caught Up with Willard Custer," Smithsonian Air & Space Magazine, April/May 2007, pp. 58-61.

[2] Englar, Robert J. and Brian A. Campbell, "Development of Pneumatic Channel Wing Powered-Lift Advanced Super-STOL Aircraft," AIAA Paper 2002-2929, AIAA 20th Applied Aerodynamics Conference, St. Louis, MO, June 25, 2002.

[3] Englar, R. J. and B. A. Campbell, "Pneumatic Channel Wing Powered-Lift Advanced Super-STOL Aircraft," AIAA Paper 2002-3275, AIAA 1st Flow Control Conference, St. Louis, MO, June 26, 2002.

[4] Englar, Robert J. & Brian. A. Campbell, "Experimental Development and Evaluation of Pneumatic Powered-Lift Super-STOL Aircraft", Paper #3 presented at the NASA/ONR Circulation Control Workshop, Hampton, VA, March 2004. Also published in "Workshop Proceedings", NASA CP 2005-213509, 2005.

[5] Englar, Robert J., "Circulation Control Pneumatic Aerodynamics: Blown Force and Moment Augmentation and Modification; Past, Present and Future," AIAA Paper 2000-2541, AIAA Fluids 2000 Meeting, Denver, CO, June 19-22, 2000.

[6] Englar, R. J. and C. A. Applegate, "Circulation Control - A Bibliography of DTNSRDC Research and Selected Outside References (Jan. 1969 through Dec. 1983)," DTNSRDC-84/052, September 1984.

[7] Englar R. J. and G.M. Blaylock, "Circulation Control Pneumatic Aerodynamics for High Lift, Flight Control, and Cruise Efficiency," Paper 6, presented at the NATO AVT-215 Innovative Control Effectors Workshop, High Lift 2 Session, Stockholm Sweden, May 20, 2013.

[8] Pugliese, A. J. (Grumman Aerospace Corporation) and R. J. Englar (DTNSRDC), "Flight Testing the Circulation Control Wing," AIAA Paper No. 79-1791, AIAA Aircraft Systems and Technology Meeting, August 1979.

[9] Englar . R. J. and Bushnell, D., "Blown Channel Wing for Thrust Deflection & Force/Moment Generation, "US Patent 7,104,498, granted Sept. 12,

2006, assignee Georgia Tech Research Corporation.

[10] Englar, R. J., "Development of Circulation Control Technology for Powered-Lift STOL Aircraft," NASA CP-2432, "Proceedings of the 1986 Circulation Control Workshop".

[11] Englar, R. J., J. H. Nichols, Jr., M. J. Harris, J. C. Eppel, and M. D. Shovlin, "Development of Pneumatic Thrust-Deflecting Powered-Lift Systems," AIAA Paper No. 86-0476, AIAA 24th Aerospace Sciences Meeting, Reno, Nevada, January 6-9, 1986.

[12] Hines, N., A. Baker, M. Cartagena, M. Largent, J. Tai, S. Qiu, N. Yiakas, J. Zentner and R. J. Englar, "Pneumatic Channel Wing Comparative Mission Analysis and Design Study, Phase I," GTRI Technical Report, Project A-5942, March 22, 2000.

[13] Eppel, J. G., M. D. Shovlin, D. N. Jaynes, R. J. Englar, J .H. Nichols Jr., "Static Investigation of the Circulation-Controlling/Upper-Surface Blowing Concept Applied to the Quiet Short-Haul Research Aircraft," NASA Technical Memorandum 84232, (July 1982).

[14] Englar, R. J., J. H. Nichols, Jr., M. J. Harris, J. C. Eppel and M. D. Shovlin, "Circulation Control Technology Applied to Propulsive High Lift Systems," SAE Paper No. 841497, SAE Aerospace Congress and Exposition, Long Beach, CA, October 15-18, 1984.

[15] Blick, E. F., "The Channel Wing- An Answer to the STOL Problem," Shell Aviation News No. 392, 2-7, 1971, June 27, 2017.

[16] Blick, Edward. F, and Vincent Homer, "Power-on Channel Wing Aerodynamics," AIAA Journal of Aircraft, Vol. 8, No. 4, pp. 234-238, April 1971.

[17] Jones, G. S., and R. J. Englar, "Advances in Pneumatic-Controlled High-Lift Systems Through Pulsed Blowing", AIAA 2003-3411, AIAA 21st Applied Aerodynamics Conference, Orlando, June 23-26, 2003.

Acknowledgements

This is the most pleasant section to write because I will relive so many memories of so many wise and generous people who have helped me over the past forty years to produce this book. Starting from the beginning of the project, I owe these people my thanks.

My recently departed mother, Sara Custer, who visited the Custer homestead to obtain and forward to me as many magazines, newspaper articles, and photographs about the Channel Wing that were available in the early 1970s. Thank you, Mom. And thank you for honestly critiquing my first draft. I have dedicated this book to you. I wish you were here to see the final product.

Joliet Custer, who sat with me for hours over many visits to the US Patent and Trademark Office's reading room where we copiously researched Willard's Channel Wing patents. I still have your handwritten notes, Joliet, listing patent number, subject and date. Thank you for your enthusiasm as we screened and then copied binders of yellowed pages on a copier that voraciously consumed one dime for each page. And thank you for later introducing me to the volumes of litigation records that form the paper trail of Willard's painful experiences in the courts.

Dr. Thomas Crouch, curator emeritus at the Smithsonian Institution's Air and Space Museum, who encouraged me to write this book. Thank you, sir, for your continuing encouragement over all these years.

Dr. F. Robert van der Linden, curator at the Smithsonian Institution's Air and Space Museum and curator of the CCW-1, who introduced me to the museum's archives and its Channel Wing documents and pictures safeguarded there by Willard Custer. Thank you for all your help and encouragement. And Miss Kate Igoe, Rights Management Archivist of the Archives Division of the Smithsonian National Air and Space Museum, who persisted through the years of the COVID-19 pandemic and museum shutdown to fill my request for scans of Channel Wing photos. Thank you for your professional kindness and effort.

My sister, Wendy Carter, who stayed in contact with the Custer clan. Thank you for sharing their love of flying and for including me in your own flying adventures. Thank you for sharing your family photos and for allowing me to use some of them in this book.

The many researchers and research librarians at the Library of Congress, the National Archives, the District of Columbia Public Library, the Johns Hopkins University Library (Kelly Spring), the Georgetown University Law Library, the Catholic University Library, Auburn University Library Archives and Special Collections (John Varner), the Carnegie Library of Pittsburgh, as well as the local libraries in Hagerstown, MD, Ventura, CA, Akron, OH, Enid, OK, McAllen, TX, and others, who meticulously searched their holdings, records, local newspapers and obituary databases for any name I submitted to them. Thank you for the gold you mined for me. I still can't believe I had to pay so little for all your help!

Fred Guerin, my FAA certification expert, who over the course of several years welcomed my emails and generously answered my questions about the FAA's Type Certification requirements and the process by which such certification might be obtained. Without you, Fred, I would not have comprehended any of it. And thank you, Bob Trainor, for leading me to him.

Chris Smith, pilot Bill Atrill's grandson, who shared with me copies of his grandfather's flight log, letters, and photographs. Thank you, Chris, for contacting me and for letting me remember your grandfather while I remembered mine.

Paulette Dorr, who also reached out to me to share her Channel Wing photographs in slides and hard copy. My collection is the richer for them.

Caroline Clark, who contacted me from Christchurch, New Zealand and sent me the letters that Walter Davidson exchanged with her father, Ernle Clark. Caroline, without your gift I never would have known about the London to Christchurch air race, the Ferry Command, or your father. Thank you for letting me remember him while I remembered my grandfather.

Dr. Edward Blick, who studied, documented, and promoted the CCW-5 from the University of Oklahoma's Aeronautics Department. Thank you, Dr. Blick, for entrusting me with your Channel Wing file. I wish you had lived to receive a complimentary copy of this book, the only thing that you requested in return.

Dr. Barnes McCormick, who responded so cordially to my inquiry regarding the test he conducted in Enid. Thank you, sir, for the added information and for copying for me the pages of your textbook that mentioned the Channel Wing.

Faye Whitehead, Rev. Bob Whitehead's widow, who opened her home to me to retell me stories of Bob and Willard and the first production model CCW-5. Thank you, Faye, for your annual Christmas card, for introducing me to your sons, Jeff and Stuart, and for the gift of the two metal Channel Wing models that Willard gave your family. Finally, thank you for the photograph of Bob and for letting me share the letters he wrote to you as he flew the CCW-5 home.

Carolyn Peat, Barr Peat's daughter, who described her memory of the CCW-2 demonstration in Pittsburgh, PA, and graciously provided a copy of her father's affidavit describing how he met Willard. Thank you, Carolyn, for teaching me more about Barr's and Willard's friendship.

James (Pittsburgh) and James (Ventura, CA) Davidson, nephew and son, respectively, of Walter Davidson, who shared with me their pictures and memories of Walt and his Ferry Command flight log. Thank you, gentlemen. Walt is a legend.

Those who read early drafts and gave me valuable advice, suggestions, and encouragement, especially, my brother Neil, Lois Willett, Paul McCulla, and Linda Sittig. Thank you, Lois, for that one question which sent me into a complete rewrite ("Do you want it to be accurate or to be remembered?"). Thank you, Paul, for your encouragement and for wanting to read more. And thank you, Linda, for your advice to use conversations to insert humor.

Those who doggedly prodded me with "How's that book coming?" especially, Phil Webster, Lynn Rayborn, Steve and Judy Sandberg, Greg and Danielle Fant, Bob and Susan Trainor, and Steve Willett. Thank you all for keeping after me to keep after it.

Bobbi Carducci, vice president of Pennwriters, Inc., and leader of Area 7, whose referrals to my excellent editors made so much difference. Thank you, also, Bobbi, for organizing the great training in the organization's Writer's Project Runway series (so aptly named for this aviation author), and for your quick answers to all my questions.

My two editors, Ramona Long and Val Muller, both kind mentors to this new author. Ramona, you passed away before this was published. Your candid questions and critique, and your thorough copy editing, wrapped in friendship and praise, made me work harder to achieve a better book. Thank you, Val, for your friendship, candid comments, advice, copy editing and encouragement. Thank you for fitting my project into your family schedule during the summer of COVID-19. You are THE BEST.

Lori Seils, my expert typesetter and new friend. Thank you, Lori, for being on my wavelength, which made working with you so easy. Your professionalism and humor made it fun to experiment to solve even the thorniest formatting problems we encountered. Thank you for putting your heart into "our" book!

Robert Englar, my recently departed co-author and friend, who jumped on the opportunity to contribute the Foreword and Afterword to this book. Thank you, Bob, for your enthusiasm, flexibility, hard work and determination, and for your ideas and encouragement. Your contributions transformed this book. I wish you were here to relish it.

Most of all, Martha Chamberlin, who pronounced my first, worst draft "a good read," which gave me courage to continue writing; who was ready to go anywhere (Akron, Pittsburgh, Auburn) my research might take us; and who shared with me some of the experts in her family for their advice and encouragement. Thank you for planning our trip to Auburn University, and for spending two days in the basement of their library, bent over a scanner, copying every photograph Frank Kelley saved in his files, and storing the images on your phone until we got them safely home. Finally, thank you for reading my "finished" manuscript, and helping me with the final polish. What a difference you have made in my book and in my life.

Finally, thank you to all those others who shared their memories and encouragement along the way, especially my father, who, in the year he died, blessed my effort with the words, "Thanks for doing this." This, Dad, is my contribution to the Custer dream. If I have done any of it well, I owe it to those named above and those others, who, because of my aging memory—not my lack of gratitude—will regrettably remain unnamed.

Notes

Chapter 1

[1]The leading quote comes from page 3 of Willard's first patent, "Aeroplane," U.S. Patent 1,708,720, 9 April 1929.

[2]Sources disagree on the year in which Willard witnessed the barn roof incident. One source sets the year at 1918 (Richard B. Weeghman, "The Queer Birds: Custer Channel Wing," Flying, April 1965, p. 43). Both *TIME* ("Science: Flying Tubes," *TIME*, 17 December 1951), and *Popular Mechanics* (Kevin Brown, "Special Report: Cockpit-Testing the Legendary Channel-Wing," *Popular Mechanics*, September 1964) agree on 1925. By the *Baltimore Sun's* (30 August 1954) calculations, it would have been 1927. I used the plurality's year.

[3]The locus of Back Creek Valley near Martinsburg, West Virginia is identified in the *TIME* story ("Science: Flying Tubes," *TIME*, 17 December 1951).

[4]The references to Glenn Curtiss and Henry Farman are taken from R. G. Grant, *Flight: The Complete History*, London: Dorling Kindersley, Ltd., 2002, p. 30 and 32, respectively.

[5]The Land Records of Washington County, Maryland, s.v. "Willard Custer," record Willard's ownership of his farm for his entire life.

[6]The short conversation between Willard and Lula on the back porch is based on my recollections of the two. Willard always had a case of the little green Coca-Cola bottles in the pantry, and Lula and he were great bridge players.

[7]The drawings (Figure 1.3) and description of Willard's first patent are taken directly from the patent as issued, Willard R. Custer, "Aeroplane," U.S. Patent 1,708,720, 9 April 1929. I added the shading. This is expired and now in the public domain.

[8]What is related of the patent application and issuance process comes from a handy pamphlet from the U.S. Patent and Trademark Office, *General Information Concerning Patents: A Brief Introduction to Patent Matters*, Washington, DC: U.S. Department of Commerce, 2005. Much of the information Garvey imparts to Willard comes from this pamphlet.

[9]The clause "a chicken-house laboratory" to describe Willard's backyard workshop comes from "The Shape of Wings to Come?" *Flying*, September 1947, p. 46. The same article describes his first company as "a symbol of small-town neighborliness and enterprise." Very quaint.

[10]The description of Willard's backyard and workshop are from my personal memory.

[11]The description of Willard's first experiments is taken from Kevin Brown, "Special Report: Cockpit-Testing the Legendary Channel-Wing," *Popular Mechanics*, September 1964.

[12]Willard's Depression-Era strategy for generating car sales by taking automobile orders and then soliciting the owners of such cars as he noticed them in traffic my father told me. He also described the family's musicality.

[13]Details of the Federal Violin Bureau are taken from *Polk's Hagerstown City Directory* (Baltimore: R.L. Polk & Company Publishers, 1937, s.v., "Federal Violin Bureau").

[14]Willard's table-top model description comes primarily from Walt Boyne, "The Custer Channel Wing Story," *Airpower*, Vol. 7 No. 3, May 1977. That article contains several excellent photographs. The photograph in Figure 1.4 is from the Custer Channel Wing Collection in the Auburn University Archives and Special Collections and was undoubtedly taken by Frank Kelley. Used by permission.

[15]The description of Frank Kelley is drawn from: 1) his obituary appearing 9 November 1996 in the Massillon Evening Independent, 2) an article about him appearing 21 August 1949 in *The Canton Repository* entitled, "Cantonian Helps Design, Test New Channel-Wing, Model T of the Air," and 3) a variety of photographs (such as Figure 1.5) and other lesser materials in the Custer Channel Wing Collection in the Auburn University Archives and Special Collections. Used by permission.

Frank's WWII studies are described in his Fun-Kell Aircraft Corporation profile, found in the Western Maryland Room, Washington County Library, Hagerstown, Maryland.

[16]John Forney Rudy ("Custer Channel Wing," *Air Trails Pictorial*, June 1948, p. 83) describes Frank Kelley's Tri-State Tank Lines as "operating petroleum motor transports."

[17]The account of Frank's and Willard's first meeting and their plan to partner in a new company is reconstructed based on subsequent events and Frank's business experience. The photograph (Figure 1.6) was obviously posed and is found in the Custer Channel Wing Collection in the Auburn University Archives and Special Collections. The signature reads, "To My Good Friend Frank Kelley."

[18]Most of what I learned about Bernard Garvey's life came from his obituary, "Bernard F. Garvey, 75, Dies; Patent Attorney for 50 Years," *Washington Post*, 22 October 1967. I saw his personality in reading the transcript of a later trial (Custer v. Ooms, Civil Action 32002, U.S. District Court for the District of Columbia, 15 May 1947). I searched high and low for a photograph with no success. Based on the court transcript, I reconstructed the conversation in Garvey's office. By the way, Garvey's office building still stands in Washington, DC.

[19]Another of the best sources for details on the early years of Willard's efforts is the Custer v. Ooms transcript. The conversation between Mr. Garvey and Willard at the end of this chapter is based on statements Garvey made at trial as recorded in the transcript.

Chapter 2

[1]The leading quote is taken from Richard B. Weeghman, "The Queer Birds: Custer Channel Wing," *Flying*, April 1965, p. 44.

[2]Again, much of what is related about Willard's early years comes from the transcript of a later trial (Custer v. Ooms, Civil Action 32002, U.S. District Court for the District of Columbia, 15 May 1947). The trip to Detroit, the referral to Reuben Davis, and Willard's subsequent relationship with Briggs Manufacturing is related in the transcript, although the company name was never revealed in court or in any publication. That name is revealed here for the very first time. In court, Garvey referred to "the Bonbright patent," but what he might have meant by it was unknown until a recent internet search returned an airplane patent by a "Howard Bonbright" (U.S. Patent 2,397,526). It was assigned to the Briggs Manufacturing Company. The patent contains word-for-word plagiarism from Willard's writings. Charles E. Sorensen (*My Forty Years with Ford*, Detroit: Wayne State Uni-

versity Press, 1956, p.265) identifies Howard Bonbright as the Treasurer of Briggs Manufacturing.

[3]The description of Briggs' dominance of the automotive manufacturing world in Detroit comes from Richard Bak, *Cobb Would Have Caught It: The Golden Age of Baseball in Detroit*, Detroit: Wayne State University Press, 1991, p. 79.

[4]The exact height and weight of Willard Custer at the time is provided by the U.S. Department of Transportation, Federal Aviation Administration, Aircraft Registry Records, NX30090, Form ACA 501, 30 July 1943.

[5]The conversation in Bonbright's office is reconstructed based largely on subsequent events and the court transcript. That Willard was thirteen when he left school is supplied by John Forney Rudy, "Custer Channel Wing," *Air Trails Pictorial*, June 1948, p. 23. Cecil Custer's background comes from his obituary, "Cecil E. Custer Dies; Retired CSC Official," (Washington) *Evening Star*, 28 October 1963.

[6]Dr. Zahm's and Dr. Crook's biographies and their relationship with The Catholic University of America, Washington, DC is from the university website, "Historic Images of The Catholic University of America: Vanished Buildings," 20 October 2003, <cuexhibits.wrlc.org/exhibits/show/vanished-buildings/biographies/crook--louis-henry--1887-1952-> and <cuexhibits.wrlc.org/exhibits/show/vanished-buildings/biographies/zahm--alfred-f---1862-1954->.

[7]That Dr. Crook was five feet tall is supplied by "Science: Happy Endings," *TIME*, 20 June 1949. His obituary appeared in "Louis Crook, C. U. Professor, Figure in Patent Suit, Dies," *Washington Post*, 20 November 1952.

[8]The initial meeting between Willard and Dr. Crook I based on my knowledge of Willard's personality. Willard was gregarious and always loved a joke. And when he joked, you could see the twinkle in his eye as he waited for the other person to "get it." Then they would both have a good laugh. At the same time, he was not afraid to say what he was thinking. There's a story of him telling a general to his face, "You don't know what you're talking about." So, while he was very respectful of people, he was not afraid of people, even powerful people. He was comfortable to say what other people would only think. In this conversation, I put the two traits (affability and transparency) together to give an example of the banter he might have started over last names. Finally, he was always hearing the same "Custer's last stand" joke, so I put it in Dr. Crook's mouth this time.

[9]The photograph of Dr. Crook at work on a model of the channel wing is found in the Custer Channel Wing Collection in the Auburn University

Archives and Special Collections. Used by permission.

[10]Much of what I relate of the patent application and appeal process comes from the U.S. Patent and Trademark Office, *General Information Concerning Patents: A Brief Introduction to Patent Matters*, Washington, DC: U.S. Department of Commerce, 2005.

[11]The record of correspondence between Willard, Garvey, and Crook, on the one side, and the patent examiners, on the other side, is preserved in what is called the file wrapper of the patent record (Willard R. Custer, "Aircraft Having High-Lift Wing Channels," U.S. Patent 2,437,684, 16 March 1948).

[12]That Frank Kelley was the source of the "Bumblebee" mascot is derived from the Custer Channel Wing Collection in the Auburn University Archives and Special Collections. He kept a photocopy of the magazine advertisement in his files. The quote itself is elsewhere attributed to Igor Sikorsky, who probably had read it in more succinct wording by the French mathematician Andre Sainte-Lague.

[13]Dr. Crook's report is extant perhaps only in the record of materials donated by Willard to the Smithsonian Institution's library (L.H. Crook, "Aeronautical Reports, Report No. 545: Preliminary Report on Custer Airplane," Bethesda (Maryland), 26 December 1941, Smithsonian Institution Archives, s.v., "Custer Channel Wing"). His affidavit for the patent examiners (filed in the U. S. Patent Office 23 May 1944) remains in the file wrapper of the contested patent.

[14]The one patent on which Willard and Crook were co-applicants is Willard R. Custer, "Airplane With High Lift Channeled Wings," U.S. Patent 2,510,959, 13 June 1950.

[15]Clear pictures of the Bumblebee in the 1940s are hard to find. Figure 2.2 is from Frank Kelley's files in the Custer Channel Wing Collection in the Auburn University Archives and Special Collections. Used by permission. Figure 2.3 is courtesy of the Smithsonian Institution's National Air and Space Museum Archives (NASM 72-9353), which records that the photo was taken at Dr. Crook's laboratory at Catholic University on 11 November 1942. Figures 2.4 (NASM A-48306-E) and 2.5 (NASM 00032782) are also courtesy of the Smithsonian Institution's National Air and Space Museum Archives. Used by permission.

[16]Details of the Bumblebee are taken primarily from "The Wing that Fooled the Experts," *Popular Mechanics*, May 1947, p. 82, and "Custer Channel Wing Plane Takes Place in Air Museum," *Hagerstown (Maryland) Daily Mail*, 22 August 1978.

[17]The Briggs Company boasted of its WWII success in a full-page advertisement appearing in *Flying*, September 1945, p.129.

[18]Willard's explanation of why the CCW-1 was called "the Bumblebee" I recall from memory. I didn't realize he referred to the aircraft itself as the Bumblebee until I showed a picture of it to my father many years later, and he excitedly exclaimed, "The Bumblebee!" That the bumblebee became a symbol of the Channel Wing's defiance of conventional aerodynamics is reflected in "Wind-Lifted Roof Recalled As Custer's Plane Hovers," *(Baltimore) Sun*, 30 August 1954.

[19]The story of the Bumblebee's maiden flight, with Willard at the helm, is told in Walt Boyne's "The Custer Channel Wing Story," *Airpower*, May 1977. All of these details also formed the basis of my reconstruction of the conversation between Willard and the Briggs representatives. That Bonbright had recently died is documented in his obituary.

[20]The FAA records of the CCW-1 can be obtained from the U.S. Department of Transportation, Federal Aviation Administration, Aircraft Registry Records, under registration number NX30090, 28 September 1943.

[21]The little airfield in Beltsville is described by Paul Freeman, on his website, "Abandoned & Little-Known Airfields: Maryland, Northern Prince Georges County," 2002, 2012, revised 15 October 2012 <www.airfields-freeman.com/MD/Airfields_MD_PG_N.htm>. The Bumblebee's testing at that field is described in the Custer v. Ooms court transcript (cited above) and by the *Popular Mechanics* article, "The Wing that Fooled the Experts," also cited above.

[22]Gen. Gilmore's visit to the airfield is related by Walt Boyne (cited above). He is further described on the website, "Wright Patterson Air Force Base History: Commanders," July 2005, at <www.ascho.wpafb.af.mil/commanders/commanders.htm>. Orville Wright's presence in Washington at that time is also documented in "Wilbur and Orville Wright, A Chronology, 1943" by the U.S. Centennial of Flight Commission, at <www.centennialofflight.net/chrono/1943.htm>.

Chapter 3

[1]Both of the leading quotes, and much of the story related in this chapter, are taken from the court transcript of Custer v. Ooms.

[2]The opening conversation is reconstructed simply to sum up for the reader the issue between Willard's patent and Henter's patent. It's a conversation that Garvey must have had with Willard at some time.

[3]Garvey's letter to the patent examiners and Dr. Crook's affidavit are in the file wrapper for the contested patent.

[4]Young's initial skepticism is reported in "The Shape of Wings to Come?" *Flying*, September 1947, pp.45 and 46.

[5]Don Young's first report is to be found as "AAF Report 5142: Test of 1/3-Scale Powered Model of Custer Channel Shaped Wing Five-Foot Wind-Tunnel, Test Number 487," Washington, DC: Army Air Forces, 5 September 1944.

[6]The synopsis of Igor Sikorski's career, to contrast with Willard's, is from the website, "Igor Sikorsky: History: Part 1," 23 January 2016, at <www.sikorskyarchives.com/History.php>.

[7]Statement of the Case, Before the Board of Appeals, US Patent Office, Appeal Number 49,908, Re: Serial Number 355,055, 6 February 1945, found in the file wrapper of the contested patent.

[8]Don Young's second report, in which he describes in detail his tests conducted at Willard's workshop, is found in Young, Donald, "Custer U-Shaped Channel Wing," Army Air Forces, Air Technical Service Command, Memorandum Report, 28 July 1945. His third, and final, report was published two years later with the quite lengthy title, "Test of Two Custer Channel Wings Having a Diameter of 37.2 Inches and Lengths of 43 and 17.5 Inches (Five-Foot Wind Tunnel Test Number 545)," Washington, DC: Army Air Forces, 14 April 1947.

[9]Garvey was kept busy. In addition to wrangling with the Patent Office over Willard's first utility patent application, he was also working the process for the several follow-on applications, including the one in concert with Dr. Crook. As the follow-ons depended on the issuance of the predecessor, their fates were similarly determined. Garvey's argument for the Custer-Crook patent is recorded in the Memorandum Supplementing Brief for Appellant, Before the Board of Appeals, US Patent Office, Appeal Number 7,602, Re: Serial Number 456,549, 12 December 1946, for Crook and Custer, "Boundary Layer Remover for Airplanes."

[10]The photograph in Figure 3.1 shows a model channel wing with a 3-blade propeller at Wright Field. This was one of the various configurations tested by Don Young. The photograph is found in the Custer Channel Wing Collection of the Auburn University Archives and Special Collections. Used by permission.

[11]The pieces of Bonbright's patent that Garvey here reads, word-for-word matching Willard's writings to the Patent Office, are taken from Howard Bonbright, "Aircraft," U.S. Patent 2,397,526, 2 April 1946.

[12]The conversation in which Garvey informs Willard of Bonbright's patent is reconstructed based on the Custer v. Ooms court transcript. That there was an agreement reached between Briggs and Willard, however, is a deduction on my part based on the circumstantial evidence. First, the identity of the Detroit sponsor has never been revealed before now. Why was it a secret? Second, Garvey must have done something when Bonbright's suspected subterfuge was discovered; he mentions it to the judge as if the Patent Office knows all about it. Third, what obligation did Willard have to Briggs for all the company's help, and why is it never mentioned? Fourth, why did Briggs' financial support suddenly stop? The entire relationship and its terms remain in the shadows. The one explanation that answers all these questions for me is that Garvey confronted the parties, leading to an agreement to sever relations and to release the parties from their obligations and liabilities, and that this agreement would itself forever remain confidential.

[13]The narrative that describes the trial of Custer v. Ooms, Civil Action Number 32002, US District Court for District of Columbia, 10 December 1945, Complaint for Issuance of United States Letters Patent, is derived largely from the court transcript. All of the dialog is verbatim, but heavily redacted. The patent diagrams (Figures 3.2, 3.3, 3.4) come from the corresponding patents, as issued, now in the public domain.

[14]The fact that Dr. Crook was still performing under contract for the National Aircraft Corporation in 1947 and 1948 is documented in checks and receipts found in the Custer Channel Wing Collection of the Auburn University Archives and Special Collections.

[15]The fact of the stipulation ending Custer v. Ooms, 26 May 1947, is found with the case documents. I've recreated the conversation at the trial's end to explain to the reader, as Garvey would have had to explain to Willard, exactly how the stipulation worked.

[16]The article, "The Shape of Wings to Come?" *Flying*, September 1947, p. 46, reports the result of the case, which is Judge Curran's stipulation that the patent be modified to say that the Channel Wing delivers vertical lift, not just superior lift.

[17]Bernard Garvey will serve Willard's stream of patent applications for the next five years. Alongside his law practice he will continue to teach at Georgetown University Law School as an adjunct professor from 1945 until 1962. When he died in 1967 after fifty years of practicing law, his obituary ("Bernard F. Garvey, 75, Dies; Patent Attorney for 50 Years," *Washington Post*, 22 October 1967) highlighted only two court cases from his long career. One was a patent infringement case that he won in 1964. The other was the patent issuance case he won for Willard Custer.

[18]The cost of legal fees for patents and related litigation cost Willard's little company $3,788.20 in 1947 and a whopping $7,121.00 in 1948, according to the company's financial statements found in the Custer Channel Wing Collection in the Auburn University Archives and Special Collections. The combined amount is equivalent to $117,000.00 in 2020 currency.

Chapter 4

[1]The first leading quote is from a letter to me from Mr. Hartman, quoted by permission. He is the inventor of the Channeled Fan Aircraft, U.S. Patent # 2994493. His patent references two of Willard's patents and employs a channel wing. The second quote is from "The Shape of Wings to Come?" *Flying*, September 1947, p. 45.

[2]Barr Peat testified in a later trial that he met Willard Custer at the Aero Club dinner in 1949 and stayed at the Custer home as a guest for three days in "Statement of D. Barr Peat Regarding Stock Option Granted by Custer Channel Wing Corporation," Draft, 9 March 1970, provided by Carolyn Peat, his daughter. Melvin Altshuler reported that Charles Lindbergh was the recipient of the Aero Club award and special speaker at the club dinner in Washington that night in his article, "Lindbergh Cites Threat to Survival," *The Washington Post*, 18 December 1949.

[3]Peat's biography is taken from *Bettis Airfield,* June 2012, <earlyaviators.com/ebettis5.htm>, and Paul Freeman, *Abandoned & Little-Known Airfields: Southwestern Pennsylvania*, 2013,<www.airfields-freeman.com/PA/Airfields_PA_SW.htm#bettis>.

[4]The signed photograph in Figure 4.1 is found in the Custer Channel Wing Collection in the Auburn University Archives and Special Collections. Used by permission. The signature reads, "To My Good Friend Frank Kelley."

[5]The testing of the CCW-2 channels, as independent demonstrations of static lift, is documented in two articles, clipped and stored without attribution in the Custer Channel Wing Collection in the Auburn University Archives and Special Collections. The picture (Figure 4.2) is a photograph given to me by Paulette Dorr and now in my collection. A copy was also found in the Custer Channel Wing Collection in the Auburn University Archives and Special Collections. Used by permission.

[6]Estimated weight of the CCW-2 varies across the documentation, but all accounts are in the neighborhood of 1,000 pounds. The plane started with the barest of bones but accrued weight over time as ailerons, etc., were added. Frank's unexpected flight in it is described in "Mr. Custer Has an Idea," *Current Science and Aviation (Railroads edition)*, Vol. XXXIV, Number

3, 27 September to 1 October 1948, p.1.

[7]There are several posed photographs of the CCW-2 with Frank, Willard or Curley in the cockpit. This signed photograph is found in the Custer Channel Wing Collection in the Auburn University Archives and Special Collections. Used by permission. The signature reads, "To My Good Friend Frank Kelley."

[8]Willard's ultimate goal, as expressed to Peat in this chapter, was published in "The Shape of Wings to Come?" *Flying*, September 1947, p. 79.

[9]Jack Heinz' biography comes from the web site, *The Heinz Family*, June 2012, <www.johnheinzlegacy.org/heinz/heinzfamily.html>.

[10]The U.S. Department of Transportation, Federal Aviation Administration, Aircraft Registry Records, are indispensable in providing timelines and statistics on each of the Channel Wing aircraft through the various government forms, some of which were filed annually. The records are accessible via each aircraft's registration number. The CCW-2's registration number was N1375V. Some of the statistics in this chapter for that aircraft are from these records.

[11]The photograph (Figure 4.4) was taken in Hagerstown and appeared in several publications. It was no doubt taken by Frank Kelley and is found in the Custer Channel Wing Collection in the Auburn University Archives and Special Collections. Used by permission.

[12]Measurements of the CCW-2 taken by Willard are documented in his whitepaper, "The Custer Channel Wing," Custer Channel Wing Corporation, Hagerstown, Maryland, 1 March 1953. The NACA study (see below) records the channel's chord to be 35 inches.

[13]C.G. Taylor's biographical information is gleaned from the following sources: *Piper*, August 2012, <www.centennialofflight.net/essay/GENERAL_AVIATION/piper/GA6.htm>; "C. Gilbert Taylor, 89, Inventor of Small Plane," *New York Times*, 12 April 1988, Obituaries; "Taylorcraft Corporation and Armour & Co.," 19 November 1956, transcript of radio broadcast by WFAH Alliance (Ohio); and Fun-Kell Aircraft Corporation Profile. The latter document is undated, but the company was only viable between July 1960 and December 1961.

[14]The Custer Channel Wing Collection in the Auburn University Archives and Special Collections contains copies of two letters penned by Taylor to Willard. Frank, always working and promoting the Channel Wing in the background, kept in contact with Taylor over the years and would establish another business arrangement with him in the 1960s.

[15]The first aircraft that Willard and Taylor agreed to develop is described in "An American Approach to the Slow Landing Problem," *The Aeroplane*, London: 11 January 1952.

[16]The Baumann Brigadier is described in *Baumann Brigadier*, December 2012, <en.wikipedia.org/wiki/Baumann_Brigadier>.

[17]That the National Aircraft Corporation's charter was amended on August 29, 1950 is stated in a letter to the stockholders dated 19 December of that year, announcing the new company name. Old stock certificates had to be swapped out. Willard signed the letter as President. A copy of it is in the Custer Channel Wing Collection in the Auburn University Archives and Special Collections. The description of Helen, the corporate Administrative Assistant and Willard's daughter, is from my memory of her.

[18]The conversation reconstructed here between Frank and C.G. Taylor is based on their continuing relationship. Frank's copies of their letters gave me a sense for Taylor's somewhat acerbic view on things. The events they discuss and the strategies behind them are based on actual events.

[19]The 6 December demonstration was not the first time the CCW-2 had been tested in this way. Existing photographs and films show at least one other occasion that occurred in a warmer season, as evidenced by leafy deciduous trees in the background and Willard's short sleeves. A picture that appears in the Pittsburgh Press (7 December 1951) is of the December demonstration, judging by the hats and coats of the participants, while a picture in London's *The Aeroplane* and some other periodicals is of the earlier, warmer occasion.

[20]This photograph (Figure 4.5) is not clear but is one my favorites because of its provenance. The photocopy was included by Walter Davidson in a letter sent to his wartime friend, Ernle Clark (see the next chapter). He clearly indicated his presence in the photo. This was a stock photo with the other important features documented, namely, Willard Custer and the limp wind sock. This photocopy, and the balance of the contents of the envelope, were given to me by Ernle Clark's daughter, Caroline. Used by permission.

[21]A copy of the letter from Dr. O'Neill to Willard is found in the Custer Channel Wing Collection in the Auburn University Archives and Special Collections.

[22]The NACA study is available from NTIS under the name NACA RM L53A09. The long name is Jerome Pasamanick, *Research Memorandum RM L53A09, Langley Full-Scale-Tunnel Tests of the Custer Channel Wing Airplane,* 7 April 1953, Washington, DC: National Advisory Committee for Aeronautics.

[23]On some of the other limitations of wind tunnel testing, see A.C. Kermide, *Mechanics of Flight*, 12th ed., Harlow (England): Pearson Education Limited, 2012, pp. 62-63.

[24]Willard found that the closer the propeller tips swept along the back edge of the channel, the greater lift he achieved. In attempts to seal the space between the two, he even went so far as to patent an idea in which little oil jets would be mounted along the back edge of the channel. These, he imagined, would spray oil into the tiny gap between the channel and the rotating propeller and thereby seal it even more against the recycling airflow. One can imagine what an oily mess over channel, propeller, and landscape that design would have made had it been implemented.

[25]David A. Anderton's articles are: "Vertical Lift Is Claimed for Channel Wing," *Aviation Week*, 17 December 1951 and "How Good Is the Custer Channel Wing?" *Aviation Week*, 15 June 1953.

[26]*Popular Mechanics*, September 1964, p. 232.

Chapter 5

[1]Richard B. Weeghman, "The Queer Birds: Custer Channel Wing," *Flying*, April 1965, p. 45.

[2]The background on the London to Christchurch Air Race comes from several sources: *Flight*, 27 February 1953, p. 246; KLM's 1953 'Bride Flight' to New Zealand subject of new movie, 7 November 2003, from the web site, <www.godutch.com/newspaper/index.php?id=482>; Airways Museum: 1953 London to Christchurch Air Race, August 2012, from the web site, <www.airwaysmuseum.com/Air race 1953 Viscount EN.htm>; and Michelle Sim, "The Last Great Air Race: London To Christchurch 1953," Air Force Museum of New Zealand blog, 10 February 2018, <www.airforcemuseum.co.nz/blog/the-last-great-air-race-london-to-christchurch-1953/>.

[3]The experience of flying in the Ferry Command is described in *Flying the Secret Sky: The Story of the Royal Air Force Ferry Command*, WGBH Specials, 12 August 2008. Gander, Newfoundland, is also the airfield where the international flights, grounded by the terror attacks in New York City and Washington, D.C., were forced to land and wait for days. Read "An Oasis of Kindness on 9/11: This Town Welcomed 6,700 Strangers Amid Terror Attacks," *USA Today*, 11 September 2017.

[4]Walter Davidson's age is documented in the *Pittsburgh Sun-Telegraph*, 29 August 1954.

[5]Ernle Clark is described in "Forgotten Flyer Gets Reserve Name,"

Christchurch (New Zealand) City Council press release, 15 May 2000. <www.scoop.co.nz/stories/CU0005/S00027/forgotten-flyer-gets-reserve-name.htm>.

[6]The photos in Figures 5.1 and 5.2 are likely taken by Frank Kelley. Figure 5.1 was given to me by Willard. Figure 5.2 photo (NASM 86-5556) is courtesy of the Smithsonian Institution's National Air and Space Museum Archives. Used by permission.

[7]Five-Seven-Charlie's maiden flight was reported by *Flight*, 31 July 1953, and "Aviation: The Channel Wing," *TIME*, 27 July 1953. The photograph (Figure 5.3) is a stock photo, probably taken by Frank Kelley, given to me by Willard.

[8]The photo of Walt Davidson is provided courtesy of his son, James. Used by permission.

[9]The initial estimate for obtaining the Type Certificate comes from two sources. Primarily, my FAA expert, Fred Guerin, offered the best-case estimate based on his knowledge of the Channel Wing and the Certification process and regulations at the time. The second source is Willard's estimate as documented in the Custer Channel Wing Corporation Stock Offering Circular, 2 August 1954, which dovetailed perfectly with Guerin's best-case scenario.

[10]The conversation between E.R. Smith, Willard and Walt is reconstructed based on the FAA records, Civil Air Regulations, and Willard's and Walt's combined naivete regarding the Type Certificate. Smith signed the air worthiness for Five-Seven-Charlie, so I have him represent the FAA, but he may not have been the actual FAA contact who delivered this information. Civil Air Regulations for each year they existed are available on the web. The ones in this chapter are taken from 1953: *Civil Air Regulations (1953), Part 3—Airplane Airworthiness—Normal, Utility, and Aerobatic Categories*. Washington, DC: Civil Aeronautics Board.

[11]Aside from his patents on the subject, Willard's interest in jet engines is published in "Channel Wing Flies Hour Over Oxnard; Inventor Says Jet Adaptation Next," *Oxnard Press Courier*, 22 January 1954, p. 3, as well as the Custer Channel Wing Corporation's annual reports cited in the text. The drawings in Figure 5.5 are from U.S. Patents 2,721,045 and 2,765,993, now in the public domain.

[12]According to CAA records, Davidson flew the CCW-5 1.5 hours in the summer of 1953, and another 10.5 hours between November 1953 and September 1954. U.S. Department of Transportation, Federal Aviation Administration, Aircraft Registry Records, N6257C.

[13]FAA regulations governing the flight parameters of the CCW-5 are recorded in U.S. Department of Transportation, Federal Aviation Administration, Aircraft Registry Records, for N6257C.

[14]"This thing is just about to break loose like a flood" comes from a letter from Willard to Frank, typed in Oxnard, and dated 18 January 1954. Frank kept it in his files found in the Custer Channel Wing Collection in the Auburn University Archives and Special Collections.

[15]Willard's income, as distinguished from that of his company, is described in Custer Channel Wing Corporation Stock Offering Circular, 2 August 1954, "The Shape of Wings to Come?" *Flying*, September 1947, pp. 46 and 79, and Richard B. Weeghman, "The Queer Birds: Custer Channel Wing," *Flying*, April 1965, p. 44.

[16]The sale of the CCW-2 and Heinz' $15,000 advance to the Corporation is found in the 1954 annual report of the Custer Channel Wing Corporation.

[17]The photograph (Figure 5.6) is enlarged from the one taken 27 August 1954 and released by the Custer Channel Wing Corporation to the news media with its news release. Stock photo now in the public domain.

[18]Geoff Sanders, "Mailbag: Channel Wing's Noise Pollution" *Aviation History*, January 2010.

[19]William Austin's eyewitness account was published four months later in *The SWATH*, December 1954, p. 4.

[20]Davidson's account of the "hovering" demonstration was published in the *Oxnard Press-Courier*, 27 August 1954, the *(Baltimore) Sun*, 28 August 1954, and the Hagerstown *Morning Herald*, 27 August 1954.

[21]A lengthier interview with Willard was published three days later in "Wind-Lifted Roof Recalled as Custer's Plane Hovers," *(Baltimore) Sun*, 30 August 1954.

[22]A copy of Walt Davidson's "Kingbird" brochure is included in the Custer Channel Wing Collection in the Auburn University Archives and Special Collections.

[23]The conversation between Walt and Frank is reconstructed based on their mutual membership in the Experimental Aircraft Association, their both being pilots, and Frank's concentrated effort to maintain good intramural and extramural communications. Walt's frustration and Frank's efforts to channel them, are based on Walt's article and Frank's personality. Frank may even have been the one to encourage Walt to write the article as he was maintaining relationships with the press.

[24]C.G. Taylor's activities during the 1950s were described in "Taylorcraft

Corporation and Armour & Co." 19 November 1956, a transcript of the radio broadcast by WFAH Alliance (Ohio).

[25]The Ray and Ray report (James G. Ray, "Flight Performances of the Custer Channel Wing CCW-5," Washington, D.C.: Ray and Ray, 21 December 1955) recorded that "Time and weather did not permit accurate measurement of other performances."

[26]The McAllen group's venture was described in "Custer Channel Wing Manufacturing Company Formed," *American Aviation Daily*, 24 September 1956, p. 148, and Corpus Christi Caller, 7 September 1956. Remarkably, true to Davidson's vision, there were no investors in the McAllen group that represented even one segment of the aviation industry.

[27]Having two different Channel Wing aircraft named "CCW-2" still confuses aviation history buffs, thanks to *Jane's All the World's Aircraft: 1956-57*, (New York: McGraw-Hill), s.v. "Custer," p. 263. By August 2, 1954, according to a corporate stock circular of that date, the original CCW-2 had been sold to an unnamed buyer for $5,000. That aircraft had been named CCW-2 because it was the second full-size Channel Wing ever built. The CCW-5 was so-named because it held five seats, hence the original naming scheme had been broken. The CCW-2 referenced by Jane's was a new two-seat design named according to the newer naming scheme.

[28]The cover photo from *Air Trails Pictorial* (June 1948) (Figure 5.7) shows an artist's rendering of a model that Willard built. While the artist's version only shows one seat, the design that the McAllen group favored seated two.

[29]Some potential markets that recommended themselves were never developed by the Custer Channel Wing Corporation, as it was not really in business to build and sell the aircraft. Consequently, these markets appeared spontaneously, usually following a series of CCW-5 demonstrations. See, for example, Rudy Arnold, "Plane Lands at 5 MPH!" *Mechanix Illustrated*, March 1958, p. 168; Kevin Brown, "Special Report: Cockpit-Testing the Legendary Channel-Wing," *Popular Mechanics*, September 1964, p. 232; and Richard B. Weeghman, "The Queer Birds: Custer Channel Wing," *Flying*, April 1965, p. 45.

[30]The conversation between Willard and Pete D'Angelo is reconstructed based on D'Angelo's memories of it as told to Joe Pappalardo in "Lunch with Willard," *Air & Space*, April-May 2007. The article relates D'Angelo's view; I have added Willard's view based on my knowledge of him.

[31]Goodyear's "Duck" is described in Jim Winchester, *The World's Worst Aircraft: From Pioneering Failures to Multimillion Dollar Disasters* (New

York: Metro Books, 2005), p.126.

[32]Willard's use of the two airports in Akron, Ohio, is noted in the FAA records, U.S. Department of Transportation, Federal Aviation Administration, Aircraft Registry Records, Form ACA-305 for N6257C, 3 April 1958.

[33]The timing of Walt Davidson's departure from the Channel Wing effort, and his feelings at the time, are my conjecture based on his background and personality. In fact, he may have left sooner. For example, his name is not mentioned in connection with the Ray and Ray test. He may have developed health problems. After the Custer camp left, he remained in California the rest of his short life and is buried there, dying in 1963 in his early fifties.

[34]Details of the tour from California to Ohio are recounted in the 1958 Annual Report of the Custer Channel Wing Corporation. Walt Boyne's review of the demonstration films was published in Walt Boyne, "The Custer Channel Wing Story," *Airpower*, May 1977, p. 18. The climax of the tour in Enid, Oklahoma is described in "Custer Channel Wing Holds Interest for Enid Residents," Enid (Oklahoma) *Daily Eagle*, 23 September 1959.

[35]The YouTube video of Willard Custer appearing on "I've Got a Secret" is found at "I've Got a Secret Custer Channelwing," 30 April 2013, <www.youtube.com/watch?v=7FhFlxbV-AU>.

Chapter 6

[1]SEC v. Custer Channel Wing Corporation, U.S. District Court for the District of Maryland, Civil Action 13500, 1962, Memorandum in Support of Defendants' Motion for Judgment, pp. 1-2.

[2]The conversation between Heinz and Willard is reconstructed to introduce the new company arrangement to the reader. Heinz certainly was behind the company setup, and he certainly briefed Willard on it. Willard's relationship to Heinz is never described anywhere, but my mother recalls hearing Willard exclaim, "Mr. Heinz" this, and "Mr. Heinz" that, as though he were quite the topic of conversation at the Custer home for a period. My sense is that Heinz directed Willard quite a bit and that Willard was uncharacteristically subservient to him, at least for the first few years. This is an indication both of Heinz' persuasiveness and Willard's humility and willingness to be guided in corporate matters for which he had no training or preparation. And don't forget that Heinz owned Five-Seven-Charlie at the time.

Just to be contrary, the very next month a blizzard clobbered the northeast United States. More than a foot of snow buried Hagerstown. In March, five weeks later, a second blizzard smacked the same area, including New York City. Pittsburgh somehow went largely unscathed. Pittsburgh's uncommon exemption from the blizzards of 1957-58 is recorded in Ben Gelber, *The Pennsylvania Weather Book*, New Brunswick: Rutgers University Press, 2002, pp. 54-56.

[3]Ownership of Five-Seven-Charlie at this point in time is recorded in U.S. Department of Transportation, Federal Aviation Administration, Aircraft Registry Records for N6257C, Bill of Sale dated 19 December 1958.

[4]Most of what is known of the particulars of the restructuring under the Custer-Frazer company is described in the 1958 Annual Report of the Custer Channel Wing Corporation. The prediction of actual aircraft production was published in "Channel Wing Flown in Demonstration," *Aviation Week*, 28 September 1959. A photo of Frazer may be found in the newspaper article, "Frazer Ditches Retirement to Form Aircraft Company," *The Morning Herald* (Hagerstown), 19 September 1958.

[5]The model CCW-10 photo (Figure 6.2) and that of Bill Atrill (Figure 6.3) are courtesy of Chris Smith, Atrill's grandson. Used by permission.

[6]The Letter to the Stockholders dated 16 August 1959, as well as the photo (Figure 6.3) is found in the Custer Channel Wing Collection in the Auburn University Archives and Special Collections.

[7]The U.S. Navy published the most detailed description of the Quantico demonstration in "A Plane With 'Copter Tricks," *Navy*, November 1959. But "Channel Wing Flown in Demonstration," *Aviation Week*, 28 September 1959, also reported it.

[8]All that is known of Channelair is documented in its one annual report, "A Report to the Stockholders," 25 November 1960, New York: Channelair, Inc. The basis of Willard's objections to Frazer's agreement with it is based on a Custer Channel Wing Corporation Statement to Stockholders, 25 April 1960, the first page (only) of which is found in the Custer Channel Wing Collection in the Auburn University Archives and Special Collections.

[9]The "Custer-Frazer v. Custer-Frazer" court battle is described in the Statement to Stockholders, 25 April 1960, mentioned above, and in Custer Channel Wing Corporation and William [sic] R. Custer, for the Benefit of Custer-Frazer Corporation, Plaintiffs, v. Joseph W. Frazer, H. Fraser Leith, John Astor Drayton and Custer-Frazer Corporation, Defendants. U.S. District Court S. D. New York, 9 December 1959.

[10]The National Aircraft Corporation's list of stockholders is dated 13 August 1947 and is found in the Custer Channel Wing Collection in the Auburn University Archives and Special Collections.

[11]Curtis Meyers, John Seburn, and Steve Christiano, in *Hagerstown, Remembering Our Aviation Heritage*, Greencastle (Penn): Historymania Publishing, 2004, p.134, identify Justin Funkhouser as the president of Victor Products.

[12]"You'd best get C.G. working on it...He's getting to be an old man" is inserted to convey Willard's sense of urgency, delight, and humor based on the circumstances. It's meant to be a joke as C.G. and Willard are the same age.

[13]The photographs of the Command-Air model (Figure 6.4) and prototype (Figure 6.5) were taken by Frank Kelley and are found in the Custer Channel Wing Collection in the Auburn University Archives and Special Collections. Used by permission.

[14]The Fun-Kell Aircraft Corporation Profile, left by Frank Kelley in a box of documents for the Washington County (Hagerstown) Library before his death, is undated, but the company was only viable between July 1960 and December 1961. It provides an inside look at the company and its principals, including, Frank, Funkhouser, and C.G. Taylor.

[15]The only record of any mishap with any Channel Wing aircraft reported in any newspaper appears in "Two Safe in Plane Crash," *Massillon (Ohio) Evening Independent*, 2 February 1962. While the 1962 Annual Report records the repairs, it never explains why they were needed. A photo of Bill Spence may be found in "Flying Barn Roof," *The Financial Post (Montreal)*, 8 November 1958, p.65.

[16]Spence's patent is W.G. Spence, "Curved Wing Structure for Aircraft," U.S. Patent 3,504,873, 7 April 1970.

[17]Richard B. Weeghman described the hangar in Hagerstown in his article, "The Queer Birds: Custer Channel Wing," *Flying,* April 1965, p. 45.

[18]The case is SEC v. Custer Channel Wing Corporation, U.S. District Court for the District of Maryland, Civil Action 13500, 1962. Note that it was a civil, not a criminal, action. Several of the parts from which this narrative is drawn are:

Attestation of Nancy Mattila, Record's Office, SEC, 19 March 1962

Letter from Arthur Doll, Small Business Administration, to W.J. Crow, SEC, 22 January 1962

Affidavit of Joseph Levy

Declaration of Trust, Stockholders Inventory and Manufacturing Trust, 28 August 1961

Memorandum in Support of Defendants' Motion for Judgment

Motion of Defendants for an Order Approving Certain Transactions

Custer Channel Wing Corporation records provided for the court

Letter to Custer Channel Wing Corporation stockholders, 26 October 1961

Memorandum of Law

Answer to Complaint for Injunction, 4 January 1962

Final Judgment

[19]Explanation of the issues involved in the case is provided by The Commonwealth Group's "Laws Surrounding Issuance of Stock," January 2013, at their web site, <commonwealthgroup.net>, as well as "Small Business Notes: Securities and Exchange Commission (SEC) Exemptions, January 2013," at <www.smallbusinessnotes.com/business-finances>. Bear in mind that SEC regulations are subject to change, and that any change often involves added complexity.

[20]In the 1970s, the Seegmiller lawfirm would sue the Custer Channel Wing Corporation for payment of attorney's fees. At that time, the corporation would be broke, so it's doubtful the firm ever got paid.

[21]The FAA's file on Fun-Kell's Channel Wing prototype can be found under U.S. Department of Transportation, Federal Aviation Administration, Aircraft Registry Records, registration number N9799C, 24 May 1961. It includes C.G. Taylor's letter to the FAA. The letter is undated but mentions "over eight years ago we built up the experimental airplane in question," i.e., N9799C, which places this letter about 1970, the same year that FAA registration records show the registration was canceled. Kurtis Meyers, Steve Christiano, and John Seburn write a little more extensively on the Fun-Kell prototype in *Hagerstown: Remembering Our Aviation Heritage*, Greencastle, PA: Historymania, 2004.

Chapter 7

[1]Richard B. Weeghman's "The Queer Birds: Custer Channel Wing," *Flying*, April 1965, p. 43.

[2]"Custer Channel Wing Day" was announced in the local newspaper on the preceding day (*The Herald Mail*, Hagerstown, Maryland, 3 July 1964). The newspaper then reported on it the following Monday morning (*The Morning Herald,* Hagerstown, Maryland, 6 July 1964, p. 13). A flyer was also

produced (Figure 7.1), containing a number of curiosities, one of which is Willard's name familiarized to "Bill."

[3]Descriptions of the production Channel Wing come from Walt Boyne, "The Custer Channel Wing Story," *Airpower,* May 1977, p. 17, note 5, and "Production Custer Channel Wing Makes Its Debut," *The Aeroplane and Commercial Aviation News*, 30 July 1964. The latter also provides the content of Willard's speech on this occasion. The photograph (Figure 7.2) is from the Smithsonian Institution's National Air and Space Museum Archives (NASM-1A32814). Used by permission.

[4]The photographs (Figures 7.3 and 7.4) are from the Smithsonian Institution's National Air and Space Museum Archives (NASM-1A32788 and NASM 97-16635, respectively). Used by permission.

[5]The text for Mr. Matthias' speech has not survived. This summary of his remarks is derived from *The Morning Herald* article, and from "Custer's Production Model Takes Bow," *Air Progress*, October/November 1964, p. 31. The content of Crossley and Firey's speeches is summarized in generalities in the news accounts. I took advantage of the circumstances to summarize to the reader all that has transpired earlier in the story.

[6]DeVore Aviation's role in the certification testing is described in *Popular Mechanics*, September 1964, p. 216, as well as "Custer Channel Wing Certification Begins," *Aviation Week & Space Technology*, 20 July 1964, p. 21, and "Production Custer Channel Wing Makes Its Debut," *The Aeroplane and Commercial Aviation News*, 30 July 1964.

[7]All that is documented of Willard's arrest is the date and that he was arrested by a U.S. Marshal. I have reconstructed it based on research into the Marshals service and my sense that it probably happened suddenly and quietly, late in the day. The living room where my grandparents watched the news after dinner, the turquoise easy chair, the afghan for warmth: these are all memories of mine. However, I never knew my grandfather had been arrested until I was doing research for this book; it was one of the many things about my grandfather I was never told. I suspect my siblings and cousins will also be surprised to learn it when they read it here.

[8]The revival of the SEC action against the Custer Channel Wing Corporation was announced in *Litigation Release No. 3369*, Securities and Exchange Commission, Washington, D.C., released Friday, November 12, 1965. New motions, exhibits, and amendments were added, such as, Motion to Amend Final Judgment, dated 12 June 1970, Exhibits "A", "B", and "C", SEC v. Custer Channel Wing Corporation, et al, Civil Action 13500, U.S. District Court for the District of Maryland.

[9]The court's criminal action against Willard is U.S.A. v. Custer Channel Wing Corporation and Willard R. Custer, Docket 2 Case No. 399, U.S. District Court for the District of Maryland, instituted 14 April 1965. The SEC still had an interest in following it, of course, and published the results in Litigation Release No. 3369, Securities and Exchange Commission, Washington, D.C., released Friday, November 12, 1965.

[10]C.G. Taylor's letter to Frank Kelley is dated 23 January 1968. Frank's response is dated 16 February. Copies of both letters are in the Custer Channel Wing Collection in the Auburn University Archives and Special Collections. C.G.'s letter to the FAA is found in U.S. Department of Transportation, Federal Aviation Administration, Aircraft Registry Records, for N5855V.

[11]Jack Kough's book is *Practical Programs for the Gifted*, Chicago: Science Research Associates, 1960. Willard's gracious letter to Dr. Kough is found in the Custer Channel Wing Collection in the Auburn University Archives and Special Collections. A copy of it was annotated by Dr. Kough and provided to Frank.

[12]Dr. Blick's publications related to the Channel Wing include Edward Blick, and Vincent Homer, "Power-on Channel Wing Aerodynamics," *Journal of Aircraft,* Vol. 8, No.4, April, 1971. His formula (Figure 7.5) appeared in Edward Blick, "The Channel Wing—An Answer to the STOL Problem?" *Shell Aviation News,* Number 392.1971, p. 2.

[13]The Figure 7.6 photograph shows Willard, in short sleeves, conferring with Dr. Blick (back to the camera) in Linden, NJ. This photograph is courtesy of Dr. Blick.

[14]FAA records for the production Channel Wing are found in U.S. Department of Transportation, Federal Aviation Administration, Aircraft Registry Records, for N5855V.

[15]The photograph in Figure 7.7 is in my collection.

[16]The Channel Wing camp's frustration with the FAA in 1970 was related to Lance Ward, "Unique Aircraft Developed Here," *Oklahoma Journal*, 26 June 1970.

[17]Desmond Allard, "D of T studies new aircraft," The Montreal Star, 22 August 1967.

[18]Edward Hudson, "Channel Wing Plane Gets Its Tryout at Teterboro," *The New York Times*, 17 March 1970.

[19]Barnes McCormick, *Aerodynamics, Aeronautics and Flight Mechanics* (New York: John Wiley and Sons, Second Edition, 1995).

[20]Dr. Blick's evaluation of the McCormick demonstration was recorded

in a newspaper article that he clipped out of the paper and filed away, and so the source will never be known.

[21]The brochure is entitled, "W.R. Custer Channeled Aircraft, Inc.," Hagerstown: W.R. Custer Channeled Aircraft, Inc., 1980. It was likely written by Bob Whitehead.

[22]The details of Bob Whitehead's experiences with flying the CCW-5 back to Pennsylvania, the letters he wrote to his wife, and the photograph at Figure 7.8, are courtesy of Faye Whitehead's generosity. Used by permission.

Epilogue

[1]Joe Pappalardo, "Lunch with Willard," *Air & Space*, May 2007.

[2]Figure 8.1 Photo taken 25 October 2003.

[3]Figure 8.2 Courtesy Custer Channel Wing Collection in the Auburn University Archives and Special Collections. Used by permission.

[4]Figure 8.3 Courtesy Custer Channel Wing Collection in the Auburn University Archives and Special Collections. Used by permission.

[5]Whereas Willard reported to the FAA that Five-Seven-Charlie had been "destroyed," Walt Boyne's word for it was "dismantled" in "The Custer Channel Wing Story," *Airpower*, May 1977.

[6]Figure 8.4 Courtesy Wendy Custer Carter. Used by permission.

[7]Figure 8.5 Courtesy Paulette Dorr. Used by permission.

[8]Figure 8.6 taken 3 June 2017.

[9]The status of the production model CCW-5 at the Mid Atlantic Air Museum is taken from an email from the president, Russell Strine, to me, dated 2 July 2013.

Bibliography

Books

Anderson, Jr., John D. *Introduction to Flight*, 7th ed. New York: McGraw-Hill, 2012.

Bak, Richard. *Cobb Would Have Caught It: The Golden Age of Baseball in Detroit.* Detroit: Wayne State University Press, 1991.

Gelber, Ben. *The Pennsylvania Weather Book*. New Brunswick: Rutgers University Press, 2002.

Grant, R. G. *Flight: The Complete History*. London: Dorling Kindersley, Ltd., 2002.

Kermide, A.C. *Mechanics of Flight*, 12th ed. Harlow (England): Pearson Education Limited, 2012.

Kough, Jack. *Practical Programs for the Gifted.* Chicago: Science Research Associates, 1960.

Meyers, Curtis, Seburn, John P., and Christiano, Steve. *Hagerstown, Remembering Our Aviation Heritage.* Greencastle (Penn): Historymania Publishing, 2004.

Polk's Hagerstown City Directory. Baltimore: R.L. Polk & Company Publishers, 1937, s.v., "Federal Violin Bureau."

Sorensen, Charles E. *My Forty Years with Ford*. Detroit: Wayne State University Press, 1956.

Winchester, Jim. *The World's Worst Aircraft: From Pioneering Failures to Multimillion Dollar Disasters*. New York: Metro Books, 2005.

Reports and Proceedings

Crook, L.H. "Aeronautical Reports, Report No. 545: Preliminary Report on Custer Airplane." Bethesda (Maryland), 26 December 1941. Smithsonian Institution Archives, s.v., "Custer Channel Wing."

Pasamanick, Jerome. *Research Memorandum RM L53A09, Langley Full-Scale-Tunnel Tests of the Custer Channel Wing Airplane*. 7 April 1953. Washington, DC: National Advisory Committee for Aeronautics, NACA RM L53A09.

Ray, James G. "Flight Performances of the Custer Channel Wing CCW-5." Washington, DC: Ray and Ray, 21 December 1955. Copy in the author's collection.

Test of 1/3-Scale Powered Model of Custer Channel Shaped Wing Five-Foot Wind-Tunnel, Test Number 487. Washington, DC: Army Air Forces, 5 September 1944. NTIS, ADA286874.

Test of Two Custer Channel Wings Having a Diameter of 37.2 Inches and Lengths of 43 and 17.5 Inches (Five-Foot Wind Tunnel Test Number 545). Washington, DC: Army Air Forces, 14 April 1947. NTIS, ADA308434.

U.S. Congress. Senate. Committee on the Judiciary. Subcommittee on Patents, Trademarks and Copyrights. *Nomination of Edwin L. Reynolds (Maryland) to be First Assistant Commissioner of Patents*. Washington, DC: GPO, 12 June 1961.

Young, Donald. *Custer U-Shaped Channel Wing*. Army Air Forces, Air Technical Service Command, Memorandum Report, 28 July 1945. NTIS, Serial Number TSEAL-S2-4586-3-2.

Articles in Journals and Magazines

"An American Approach to the Slow Landing Problem." *The Aeroplane*, 11 January 1952.

Anderton, David A. "How Good Is the Custer Channel Wing?" *Aviation Week*, 15 June 1953.

________. "Letters: On Custer, No Bias." *Aviation Week*, 7 September 1953.

________. "Vertical Lift Is Claimed for Channel Wing." *Aviation Week*, 17 December 1951.

Arnold, Rudy. "Plane Lands at 5 MPH!" *Mechanix Illustrated*, March 1958, 83.

Austin, William D. Editorial. *The SWATH*, December 1954.

"Aviation: The Channel Wing." *TIME*, 27 July 1953.

Blick, Edward E. "The Channel Wing—An Answer to the STOL Problem?" *Shell Aviation News*, Number 392.1971.

Blick, Edward F. and Homer, Vincent. "Power-on Channel Wing Aerodynamics." *Journal of Aircraft*, Vol. 8, No.4, April, 1971.

Boyne, Walt. "The Custer Channel Wing Story." *Airpower* Vol. 7 No. 3 (May 1977).

Briggs Manufacturing Company advertisement in *Flying*, September 1945.

Brown, Kevin. "Special Report: Cockpit-Testing the Legendary Channel-Wing." *Popular Mechanics*, September 1964.

"Channel Wing Flown in Demonstration." *Aviation Week*, 28 September 1959.

"Custer's Production Model Makes Bow." *Air Progress*, October/November 1964.

"Custer Channel Wing Certification Begins." *Aviation Week & Space Technology*, 20 July 1964.

"Custer Channel Wing Manufacturing Company Formed." *American Aviation Daily*, 24 September 1956.

Davidson, Walter J. "A Thought on Aviation..." *Experimenter*, January 1955.

Flight. 27 February 1953.

Flight. 3 July 1953.

Flight. 31 July 1953.

Flying. October 1947.

Flying. November 1947.

Flying. January 1954.

Guttman, Robert. "Custer's Channel Wing." *Aviation History*, July 2009.

Kohn, Leo J. "The Channel Wing." *Experimenter,* March 1955.

Mitchell, Kent A. "Mr. Custer and His Channel Wing Airplanes." *Journal of the American Aviation Historical Society* (Spring 1993).

Pappalardo, Joe. "Lunch With Willard." *Air & Space*, April-May 2007.

"A Plane With 'Copter Tricks." *Navy*, November 1959.

Rudy, John Forney. "Custer Channel Wing." *Air Trails Pictorial*, June 1948.

Sanders, Geoff. "Mailbag: Channel Wing's Noise Pollution." *Aviation History*, January 2010.

"Science: Flying Tubes." *TIME*, 17 December 1951.

"Science: Happy Endings." *TIME*, 20 June 1949.

"The Shape of Wings to Come?" *Flying*, September 1947.

Weeghman, Richard B. "The Queer Birds: Custer Channel Wing." *Flying*, April 1965.

"The Wing that Fooled the Experts." *Popular Mechanics*, May 1947.

Articles in Encyclopedias

Jane's All the World's Aircraft: 1956-57. New York: McGraw-Hill, s.v. "Custer."

Articles in Newspapers

"4,000 Persons See Channel Wing Demonstration Here; Matthias Lauds Builder for Perseverance." *Hagerstown (Maryland) Morning Herald*, 6 July 1964.

"A-C Airport Seeks to Expand Service." *Massillon (Ohio) Evening Independent*, 19 November 1953.

"Aircraft Plant To Operate Here." *McAllen (Texas) Valley Evening Monitor*, 17 September 1956.

"An Oasis of Kindness on 9/11: This Town Welcomed 6,700 Strangers Amid Terror Attacks." *USA Today*, 11 September, 2017.

Allard, Desmond. "D of T studies new aircraft." *The Montreal Star*, 22 August 1967.

Altshuler, Melvin. "Lindbergh Cites Threat to Survival." *The Washington Post*, 18 December 1949.

"Axis Sally." *Charleston Gazette*, 26 March 1949.

"Backyard Inventor's Plane Can Hover At Only 11 M.P.H." *(Baltimore) Sun*, 28 August 1954.

"Barrel-Shaped Wings Slow Plane to 11 MPH." *Philadelphia Inquirer*, 28 August 1954.

"Bernard F. Garvey, 75, Dies; Patent Attorney for 50 Years." *Washington Post*, 22 October 1967.

"Blick Studying Channel Wing Design Concept." A clipping from Dr. Blick's file and in the author's collection, without a periodical name or date.

"C. Gilbert Taylor, 89, Inventor of Small Plane." *New York Times*, 12 April 1988, Obituaries.

"Cecil E. Custer Dies; Retired CSC Official." *(Washington) Evening Star*, 28 October 1963.

"Channel Wing Craft Sets Another Mark." *Hagerstown (Maryland) Morning Herald*, 27 August 1954.

"Channel Wing Flies Hour Over Oxnard; Inventor Says Jet Adaptation Next." *Oxnard Press Courier*, 22 January 1954.

"Channel Wing July Fourth Celebration Set Tomorrow." *Hagerstown (Maryland) Herald Mail*, 3 July 1964.

"Channel Wing Plane Hovers At Airport." *Oxnard (Calif.) Press-Courier*, 27 August 1954.

"Custer Channel Wing Holds Interest for Enid Residents." *Enid (Oklahoma) Daily Eagle*, 23 September 1959.

"Custer Channel Wing Plane Takes Place in Air Museum." *Hagerstown (Maryland) Daily Mail*, 22 August 1978.f

"Even Tougher to Go Slower." *Philadelphia Inquirer*, 29 August 1954.

"Frank Kelley." *Massillon (Ohio) Evening Independent*, 9 November 1996, Obituaries.

"Hagerstown Man Invents 'Channel Wing,' New Type Airplane Wing; Will Change Plane Performance." *Hagerstown (Maryland) Daily Mail*, 27 April 1947.

Hudson, Edward. "Channel Wing Plane Gets Its Tryout at Teterboro." *New York Times*, 17 March 1970.

"Jack Kough." *Morgan Hill (California) Times*, 15 December 1982.

"Louis Crook, C. U. Professor, Figure in Patent Suit, Dies." *Washington Post*, 20 November 1952.

"McAllen Plane in Flight." *Corpus Christi Caller*, 22 September 1956.

"Mr. Custer Has an Idea." *Current Science and Aviation (Railroads edition)*, Vol. XXXIV, Number 3, 27 September to 1 October 1948.

"New Theory in Aviation Tested Here." *Oxnard Press Courier*, 16 June 1953.

"New Wing Aids Plane In 'Vertical' Take-off." *Pittsburgh Sun-Telegraph*, 29 August 1954.

"Plane 'Sucked Up' At 3000 Ft. a Minute, Lands at 11 MPH." *Los Angeles Examiner,* 28 August 1954.

"Plane With Novel 'Channel Wings' Ascends at Only 25 Miles an Hour." *Washington Post-Times Herald*, 19 September 1959.

Price, Mark J. "Local history: Akron's inflatable airplane is oddity of sky in 1950s." *Akron Beacon Journal*, 18 June 2012.

"Production Custer Channel Wing Makes Its Debut." T*he Aeroplane and Commercial Aviation News,* 30 July 1964.

"Scoop-Winged Plane Flies 11 MPH at Oxnard Airport." *Ventura (California) County Star-Free Press*, 28 August 1954.

"SEC Halts Custer Corp. Stock Sales." *Hagerstown (Maryland) Morning Herald,* 27 December 1960.

"Slow-Speed Airplane Is Developed." *Charleston (South Carolina) News and Courier*, 28 August 1954.

"Something New Under the Sun." *Pittsburgh Press*, 7 December 1951.

"'Tornado' Plane Hovers at 11 MPH." *New York Herald Tribune*, 29 August 1954.

"Two Safe in Plane Crash." *Massillon (Ohio) Evening Independent*, 2 February 1962.

"Unusual Airplane Demonstrated." *Pittsburgh Press*, 6 December 1951.

"Walter Davidson." *Ventura County Star Free Press (Obituaries)*, 24 August 1963.

Ward, Lance. "Unique Aircraft Developed Here." *Oklahoma Journal*, 26 June 1970.

"Wind-Lifted Roof Recalled As Custer's Plane Hovers." *(Baltimore) Sun*, 30 August 1954.

Unpublished Materials

Atrill, William. 26/27 April 1977. Letter to Chris Smith, photocopy in the author's collection.

________. 10 October 2000. Letter to Chris Smith, photocopy in the author's collection.

________. 1959 – 1960. Royal Canadian Air Force Pilot's Flying Log Book, photocopy in the author's collection.

Custer Channel Wing Corporation, corporate documents

"Annual Report 1958 of the Custer Channel Wing Corporation." Hagerstown (Maryland), photocopy in the author's collection.

"Annual Report 1962 of the Custer Channel Wing Corporation." Hagerstown (Maryland), photocopy in the author's collection.

"Annual Report 1963 of the Custer Channel Wing Corporation." Hagerstown (Maryland), photocopy in the author's collection.

Custer Channel Wing Corporation Stock Offering Circular, 2 August 1954, photocopy in the author's collection.

Custer, Willard R. 15 November 1962. Letter to Stockholders of the Custer Channel Wing Corporation, photocopy in the author's collection.

________. 20 December 1963. Letter to Stockholders of the Custer Channel Wing Corporation, photocopy in the author's collection.

________. "The Custer Channel Wing." 1 March 1953. Custer Channel Wing Corporation, Hagerstown, Maryland, photocopy in the author's collection. It was later repackaged and entitled, "How to Fly Slow and Hover With Fixed Wing Aircraft," by Willard. R. Custer.

________. "Theory of Channel Wing Aircraft: Speed of Air." Undated. Custer Channel Wing Corporation, Hagerstown, Maryland, original copy in the author's collection.

Custer Channel Wing Collection, Auburn University Archives and Special Collections, Record Group 187, Accession Number 04-040, 1920-1964.

Davidson, Walter J. 10 September 1953. Letter to Ernle Clark, original in the author's collection.

Frank Kelley. 22 December 1967. Letter to "Dick," photocopy in the Western Maryland Room, Washington County Library, Hagerstown, Maryland.

Fun-Kell Aircraft Corporation Profile. July 1960 to December 1961. Original at Western Maryland Room, Washington County Library, Hagerstown, Maryland.

Hartman, Thomas A. 4 August 2003. Letter to the author, in the author's collection.

Land Records of Washington County, Maryland, s.v. "Willard Custer."

Mathias, Charles McCurdy, Jr. Papers 1940-1990. Johns Hopkins University Milton S. Eisenhower Library, Baltimore, MD.

"A Report to the Stockholders." 25 November 1960. New York: Channelair, Inc., photocopy in the author's collection.

Smith, Chris. 7 October 2011. Letter to the author, in the author's collection.

"Statement of D. Barr Peat Regarding Stock Option Granted by Custer Channel Wing Corporation," Draft. 9 March 1970. (Courtesy of Carolyn Peat)

"Taylorcraft Corporation and Armour & Co." 19 November 1956. Transcript of radio broadcast by WFAH Alliance (Ohio). Available at <taylorcraft.org/docs/taylorcraft_wfah_1956.pdf>.

"W.R. Custer Channeled Aircraft, Inc." Hagerstown: W.R. Custer Channeled Aircraft, Inc., 1980. Probably authored by Robert O. Whitehead, president.

Legal Cases

Custer v. Ooms. Civil Action 32002. U.S. District Court for the District of Columbia, 15 May 1947. Available through the National Archives.

Custer Channel Wing Corporation and William [sic] R. Custer, for the Benefit of Custer-Frazer Corporation, Plaintiffs, v. Joseph W. Frazer, H. Fraser Leith, John Astor Drayton and Custer-Frazer Corporation, Defendants. U.S. District Court S. D. New York, 9 December 1959.

Custer Channel Wing Corp v. U S. U.S. Supreme Court Transcript of Record with Supported Pleadings. Vincent L. Gingerich, Philip A. Loomis, Vincent L. Gingerich. In the series, The Making of Modern Law, U.S. Supreme Court Records and Briefs, 1832-1978.

SEC v. Custer Channel Wing Corporation. U.S. District Court for the District of Maryland. Civil Action 13500. 1962. Available through the National Archives.

Willard R. Custer and Custer Channel Wing Corporation v. The UNITED STATES. No. 181-78. 14 May 1980. United States Court of Claims. Available through the Unites States Court of Federal Claims.

Video Recordings

Custer Channel Wing. Harold Custer: 2000, copy in the author's collection.

Flying the Secret Sky: The Story of the Royal Air Force Ferry Command. WGBH Specials, 12 August 2008.

"I've Got a Secret Custer Channelwing." 30 April 2013. <www.youtube.com/watch?v=7FhFlxbV-AU>.

United States Government Documents

Airworthiness Certification of Aircraft and Related Products. Washington, DC: Department of Transportation, Federal Aviation Administration. Order 8130.2G. 31 August 2010.

Civil Air Regulations (1953), Part 3—Airplane Airworthiness—Normal, Utility, and Aerobatic Categories. Washington, DC: Civil Aeronautics

Board.

Crook, Louis H., and Custer, Willard R. Boundary Layer Remover for Airplanes. U.S. Patent 2,428,737, 7 October 1947.

Bonbright, Howard. *Aircraft*. U.S. Patent 2,397,526, 2 April 1946.

Custer, Willard R. Aeroplane. U.S. Patent 1,708,720, 9 April 1929.

________. Aircraft. U.S. Patent 1,868,832, 26 July 1932.

________. Airplane. U.S. Patent 2,194,596, 26 March 1940.

________. Aircraft Having High-Lift Wing Channels. U.S. Patent 2,437,684, 16 March 1948.

________. Airplane With High Lift Channeled Wings. U.S. Patent 2,510,959,13 June 1950.

________. Multiple Propeller Wing Channel. U.S. Patent 2,424,556, 29 July 1947.

General Information Concerning Patents: A Brief Introduction to Patent Matters. Washington, DC: U.S. Department of Commerce, 2005.

Hartman, Thomas A. Channelled Fan Aircraft. U.S. Patent 2,994,493, 1 August 1961.

Litigation Release No. 2166. 27 December 1961. Washington, DC: Securities and Exchange Commission.

Release No. 4311. 30 December 1960. Washington, DC: Securities and Exchange Commission.

Spence, W. G. Curved Wing Structure for Aircraft. U.S. Patent 3,504,873, 7 April 1970.

Taylor, Clarence Gilbert. Lift-propulsion Device for Aircraft. U.S. Patent 2,961,188, 22 November 1960.

U.S. Department of Transportation, Federal Aviation Administration, Air-

craft Registry Records, available from aircraft.faa.gov/e.gov/ND/, for the following aircraft.
NX30090 (CCW-1)
N1375V (CCW-2)
N6257C (CCW-5 prototype)
N5855V (CCW-5 production model)
N9799C (Command-Air)

Web pages

Airways Museum: 1953 London to Christchurch Air Race. August 2012. <www.airwaysmuseum.com/Air race 1953 Viscount EN.htm>.

Aviation Glossary: Type Certificates. August 2012. <aviationglossary.com/?s=type+certificate/>.

Bettis Airfield. June 2012. <earlyaviators.com/ebettis5.htm>.

Baumann Brigadier. December 2012. <en.wikipedia.org/wiki/Baumann_Brigadier>.

The *Commonwealth Group: Laws Surrounding Issuance of Stock*. January 2013. <commonwealthgroup.net>.

Edward Curran. March 2013. <judgepedia.org/index.php/Edward_Curran>.

Freeman, Paul. *Abandoned & Little-Known Airfields: Maryland, Northern Prince Georges County*. 2002, 2012, revised 15 October 2012. <www.airfields-freeman.com/MD/Airfields_MD_PG_N.htm>.

________. *Abandoned & Little-Known Airfields: Southwestern Pennsylvania*. 2013. <www.airfields-freeman.com/PA/Airfields_PA_SW.htm#bettis>.

The Heinz Family. June 2012. <www.johnheinzlegacy.org/heinz/heinzfamily.html>.

Historic Images of The Catholic University of America: Vanished Buildings:

Alfred Zahm Biography. 20 October 2003.
<cuexhibits.wrlc.org/exhibits/show/vanished-buildings/biographies/zahm--alfred-f---1862-1954->.

Historic Images of The Catholic University of America: Vanished Buildings: Louis Crook Biography. 20 October 2003.
<cuexhibits.wrlc.org/exhibits/show/vanished-buildings/biographies/crook--louis-henry--1887-1952->.

Historic Images of The Catholic University of America: Vanished Buildings: Wind Tunnel, Catholic University Aeronautical Laboratory, Stucco Building. 20 October 2003.
<cuexhibits.wrlc.org/exhibits/show/vanished-buildings/buildings/wind-tunnel---catholic-univers>.

Hollywood Ten Trials: 1948-1950. March 2013.
<www.encyclopedia.com/doc/1G2-3498200186.html>.

How Much Cars Cost In The 60's. January 2013.
<www.thepeoplehistory.com/60scars.html>.

Igor Sikorsky: History: Part 1. 23 January 2016.
<www.sikorskyarchives.com/History.php>

KLM's 1953 'Bride Flight' to New Zealand subject of new movie. 7 November 2003.
<www.godutch.com/newspaper/index.php?id=482>.

The Lycoming Museum. March 2013.
<www.lycoming.textron.com/company/pdfs/Lycoming-Museum-Brochure.pdf>.

Piper. August 2018.
<www.centennialofflight.net/essay/GENERAL_AVIATION/piper/GA6.htm>.

Rumerman, Judy. *The National Advisory Committee for Aeronautics (NACA)*. 11 August 2012.
<www.centennialofflight.net/essay/Evolution_of_Technology/NACA/Tech1.htm>.

Small Business Notes: Securities and Exchange Commission (SEC) Exemptions. January 2013.
<www.smallbusinessnotes.com/business-finances>.

Taylorcraft. August 2012.
<www.pilotfriend.com/aircraft%20performance/Taylorcraft.htm>.

Type Certificate. August 2018.
<en.wikipedia.org/wiki/Type_certificate>.

U.S. Small Business Administration: Venture Capital. August 2018.
<www.sba.gov/business-guide/plan-your-business/fund-your-business>.

Wilbur and Orville Wright, A Chronology, 1943. U.S. Centennial of Flight Commission.
<www.centennialofflight.net/chrono/1943.htm>.

Wright Patterson Air Force Base History: Commanders. July 2005.
<www.ascho.wpafb.af.mil/commanders/commanders.htm>.

Index

Milestones

CCW-1 Maiden Flight 28

CCW-2 Demonstration 191

CCW-2 Maiden Flight 62

CCW-5 Gear-Up Landing 109

CCW-5 "Hover" Demonstration 90–92

CCW-5 Maiden Flight 81

Cimmaron Field Demonstration 146

Custer Channel Wing Day 130, 135–137, 148, 158, 211

Quantico Marine Base Demonstration 107, 110, 115, 127, 134, 209

Teterboro Demonstration 145

Organizations

Air Force Air Materiel Command 33

Briggs Manufacturing Company 17, 25, 27, 31, 33, 41, 44–45, 60, 97, 104, 153, 195–196, 217

CAA/FAA iii–iv, xiv, 30–31, 72, 83–86, 88, 90, 99–101, 104, 106–107, 113–114, 117, 120, 125–126, 129, 134, 136–137, 141, 142, 144, 146, 148–149, 151–153, 155–159, 161–162, 186, 190, 198, 205, 208, 211, 213–214

Catholic University 20–21, 72, 153, 190, 196–197, 226–227

Channelair 111–114, 116, 209, 223

Corporation Trust Company 136

Custer Channel Wing, Canada 107

Custer Channel Wing Corporation 67, 73–74, 83, 86, 88, 101, 104–105, 113–115, 117, 119, 121–124, 140, 145, 147–148, 158–159, 201–202, 205–213, 222–224

Custer Channel Wing Manufacturing Company 95, 207, 218

Custer-Frazer 104–105, 112–114, 116–117, 121, 123, 209, 224

DeVore Aviation 134, 143, 158

Federal Violin Bureau 11, 194, 215

Fun-Kell 116–117, 122, 126, 140, 194, 202, 210–211, 223

Goodyear 97–99, 101, 107

Kerr Aviation Services 146

NACA/NASA v, xi, 72, 74–76, 83, 88, 95, 98, 111, 115, 142, 155, 162–163, 165, 167–170, 181, 186–188, 202–203, 216, 227, 234

National Aircraft Corporation 13, 18, 68, 115, 200, 203, 209

SEC xiv, xvi, 117–123, 125–126, 135–139, 210–213, 221, 224, 228

Securities and Exchange Commission 115, 117, 119, 211, 212–213, 225, 228

Taylorcraft 66, 70–71, 84, 94, 97, 116, 202, 206, 223, 228

US Army Air Force 35

US Patent Office 15, 21–23, 34, 35–36, 39, 41, 43–49, 52–53, 57, 60, 75, 99, 197, 199, 200

W.R. Custer Channeled Aircraft 147, 159, 214, 223

People

Atrill, William “Bill” 107–111, 114–115

Baumann, Jack 67–68, 77, 88

Blick, Edward F. 142–146

Bonbright, Howard 18–19, 20, 25, 27–28, 44–45, 113, 195–196, 198–199, 200, 225

Boyne, Walter “Walt” xiii, xiv, 73, 100, 131–132, 194, 198, 208, 212, 214, 217

Brooks, Ward 117, 125

Clark, Ernle 79, 82, 190, 203–204, 223

Crook, Louis 20–26, 31–33, 36–39, 46, 53–54, 60, 72, 142, 153, 196–197, 199–200, 216, 220, 225, 227

Crouch, Thomas “Tom” xiii, xv, 189

Curran, Edward 45–46, 48–49, 50, 52, 54–56, 60, 200, 226

Custer, Cecil 3, 19, 89, 112, 122, 138–139, 196, 220

Custer, Harold “Curley” 5, 11, 15, 26, 42, 61–62, 64–65, 78, 100, 123, 135, 145, 202, 224

Custer, Helen 5, 28, 42–43, 68, 103–104, 113, 135, 138, 151, 203

Custer, Kenneth "Reed" xv, 10–11, 26, 37, 42, 90, 135

Custer, Lula 5–6, 8–9, 11, 13–14, 37, 42, 88, 101, 132, 135, 137–139, 151, 193

Custer, Vivian 3, 5, 42, 135

D'Anna, Pete 97, 101

Davidson, Walt 69, 71, 78, 83, 92–93, 100, 107, 115, 190–191, 203–208, 218, 222–223

Davis, Albert 93, 95, 97

Davis, Reuben 19–21, 25–26, 28, 30, 33, 195

Frazer, Joseph 105, 107, 111–114, 209, 224

Garvey, Bernard 14–15, 22–24, 35–36, 39–41, 43–50, 52–54, 57, 194–195, 197–200, 219

Gilmore, William 33, 37, 41, 69, 72, 153, 198

Heinz II, H. Jack 63–64, 67–69, 77, 83–84, 89, 93, 103–105, 113–114, 136, 156, 202, 206, 208, 226

Henter, Matthias 35, 44, 46–51, 54, 57, 198

Kelley, Frank 12–14, 18, 24, 25, 31, 33, 42, 46, 62, 66, 72, 88, 93, 100, 115, 122, 140, 192, 194–195, 197, 201–202, 205, 210, 213, 220, 223

Kough, Jack 141, 146, 213, 215, 220

Leone, Louis 95–97

Marshall, Thurgood 139

Matthias, "Mac" 35, 132, 134, 212, 219

McCormick, Barnes 146–147, 191, 213

Peat, D. Barr 59–60, 62–64, 69–70, 76, 104, 112–113, 191, 201–202, 223

Reynolds, Edwin 35, 45, 48–52, 54, 56–57, 186, 216

Sikorsky, Igor 38–39, 197, 199, 227

Spence, William "Bill" 105, 107, 114, 123–124, 145, 210, 225

Stoner, Sam 101, 141–143, 146

Taylor, C. Gilbert 64–68, 84, 88, 94, 99, 100, 115, 117, 135, 140, 151, 155, 202–203, 206, 210–211, 213, 220, 225

Wallace, Bruce 144, 147, 148

Whitehead, Robert "Bob" 147–149, 159, 191, 214, 223

Winter, Harris 125–126, 138–139, 141, 147

Wright, Orville 3, 20, 33, 59, 94, 134, 198, 228

Young, Donald "Don" 35, 37, 40–42, 53–57, 60–61, 74, 153, 155, 199, 216

Zahm, Alfred 20–21, 27, 196, 227

Places

Akron, OH 97–102, 107–108, 110–111, 114, 116–117, 123, 142, 190, 192, 208, 221

Beltsville, MD 31–33, 46, 49, 153, 198

Enid, OK 101, 141–142, 146, 158, 190–191, 208, 220

Hagerstown, MD 1, 4–5, 11–13, 18, 20, 30, 38, 40, 42, 60, 62, 64, 66, 68, 72, 82, 90, 92–93, 100, 104–105, 107, 110, 112–117, 120, 123–126, 129–131, 133, 138, 151, 158, 190, 194–195, 197, 202, 206, 208, 209–211, 214, 215, 219–223

Linden, NJ 143, 189, 213

McAllen, TX 95–97, 99, 101, 106, 113, 117, 133, 190, 207, 219, 221

Montreal, Canada 78, 105, 107, 112–114, 123, 133, 210, 213, 219

Oxnard, CA 78–79, 89, 92–93, 116–117, 140, 156, 205–206, 220–221

Pittsburgh, PA 59–60, 63–64, 66, 69–70, 78, 88, 93, 104, 113–114, 156, 190–192, 203–204, 209, 221

Shreveport, LA 100

Wright Field, OH 33–34, 37, 40–43, 53–54, 72, 98, 133, 153, 155, 199

About the Authors

Joel C. Custer

Joel Custer was born in Baltimore, MD. A graduate of Asbury University (Wilmore, KY), he holds degrees in Divinity (M.Div., Trinity Evangelical Divinity School, Deerfield, IL), Software Systems Engineering (M.S., George Mason University, Fairfax, VA), and Computer Security and Information Assurance (The George Washington University, Washington, DC), supplemented by project management and information security professional certifications. His information systems career has spanned the mortgage and telecommunications industries, as well as the federal government, including the Departments of Justice, Homeland Security, and Defense. He is the author of numerous technical documents detailing system requirements, operational concepts, designs, and test plans. Aside from the Custer Channel Wing, his hobbies include George Armstrong Custer, Biblical languages and literature, travel, and music. He lives in Ashburn, VA.

Robert J. Englar (1944 – 2021)

Robert J. Englar was a Principal Research Engineer at the Georgia Tech Research Institute (GTRI) and had over 44 years of experience in advanced aerodynamic research (much of it Circulation Control Aerodynamics related), advanced concept development, and experimental techniques. As Principal Research Engineer in the Aerospace & Acoustics Technologies Division of GTRI's Aerospace, Transportation & Advanced Systems Laboratory, Mr. Englar was responsible for research project direction and development of advanced technologies in aerodynamics. After conducting aero research during employment with the US Navy and Lockheed Martin, he originated, conducted and/or directed over 134 such sponsored research programs at GTRI since January 1989. As an example, he was Principal Investigator leading NASA-sponsored programs on Circulation Control Aerodynamics and pneumatic Powered Lift Aircraft. He holds 11 patents and 19 invention disclosures on these advanced concepts and published 209 papers and technical reports on these and other research projects. His specialty was combining this advanced aerodynamic research and development experience into design, application and verification of Circulation Control and pneumatic high-lift and 3-axis control systems which can be employed on efficient Short Takeoff & Landing (STOL and Extreme STOL) and high-performance aircraft. He was elected as a GTRI Technical Fellow based on his involvement with advanced aerodynamics and related advanced concepts. He also taught Senior design courses in the Georgia Tech Aerospace Engineering and Mechanical Engineering Schools. He was an Associate Fellow of the American Institute of Aeronautics and Astronautics (AIAA) and was a founding member of the original AIAA Applied Aerodynamics Technical Committee. He was elected AIAA Young Engineer/Scientist of the Year for his work on Circulation Control, received numerous professional awards, and lectured on pneumatic powered lift and advanced concepts to government agencies, technical societies, private industry and universities.

R. J. Englar retired from GTRI in January 2013, but acted as Advanced Aerodynamics Consultant until his death, particularly in the development of advanced pneumatic aerodynamic technologies for ground- and water-based vehicles (some commercial or high-performance racing) as well as for advanced STOL and high-performance aircraft.

Made in the USA
Columbia, SC
24 September 2023

07e77bdc-2a82-46e9-8324-ea19f7805976R01